# CONCENTRATE MANAGEMENT IN DESALINATION

## CASE STUDIES

EDITED BY
Conrad G. Keyes Jr., Sc.D., P.E., L.S., D.WRE
Michael P. Fahy, P.E.
Berrin Tansel, Ph.D., P.E., D.WRE

PREPARED BY
Task Committee on Development of Prestandards for Concentrate Management Case Studies
Desalination and Water Reuse Committee
Water, Wastewater, & Stormwater Council
Environmental and Water Resources Institute

Published by the American Society of Civil Engineers

Cataloging-in-Publication Data on file with the Library of Congress.

Published by American Society of Civil Engineers
1801 Alexander Bell Drive
Reston, Virginia, 20191-4400
www.asce.org/pubs

Any statements expressed in these materials are those of the individual authors and do not necessarily represent the views of ASCE, which takes no responsibility for any statement made herein. No reference made in this publication to any specific method, product, process, or service constitutes or implies an endorsement, recommendation, or warranty thereof by ASCE. The materials are for general information only and do not represent a standard of ASCE, nor are they intended as a reference in purchase specifications, contracts, regulations, statutes, or any other legal document. ASCE makes no representation or warranty of any kind, whether express or implied, concerning the accuracy, completeness, suitability, or utility of any information, apparatus, product, or process discussed in this publication, and assumes no liability therefore. This information should not be used without first securing competent advice with respect to its suitability for any general or specific application. Anyone utilizing this information assumes all liability arising from such use, including but not limited to infringement of any patent or patents.

ASCE and American Society of Civil Engineers—Registered in U.S. Patent and Trademark Office.

*Photocopies and permissions.* Permission to photocopy or reproduce material from ASCE publications can be obtained by sending an e-mail to permissions@asce.org or by locating a title in ASCE's online database (http://cedb.asce.org) and using the "Permission to Reuse" link.

Copyright © 2012 by the American Society of Civil Engineers.
All Rights Reserved.
ISBN 978-0-7844-1210-7 (paper)
ISBN 978-0-7844-7678-9 (e-book)
Manufactured in the United States of America.

# Contents

| | | |
|---|---|---|
| **Chapter 1** | **Introduction** ............................................................................. 1 | |
| | *Conrad G. Keyes, Jr., Michael P. Fahy, and Kenneth Mercer* | |
| | Background of EWRI Work ............................................................... 1 | |
| | Case Study Topical Areas .................................................................. 2 | |
| | Scope of the Monograph ................................................................... 3 | |
| **Chapter 2** | **Overview of Desalination Processes and Configurations** .......... 5 | |
| | *Sandeep Sethi, Val S. Frenkel, and Kenneth Mercer* | |
| | Introduction ....................................................................................... 5 | |
| | Membrane Desalination Processes .................................................... 5 | |
| | Characteristics of Concentrate Stream ............................................. 8 | |
| | Types of Concentrate Management Methods ................................. 9 | |
| | Concentrate Management and Disposal Methods ........................ 10 | |
| | Pilot Testing of Concentrate Minimization Methods .................. 13 | |
| **Chapter 3** | **Regulation of Concentrate Management** .................................. 17 | |
| | *Kenneth Mercer* | |
| | Concentrate Management Regulation Overview ......................... 17 | |
| | National Pollutant Discharge Elimination System ...................... 17 | |
| | Residuals Treatment Guidelines for Discharge to Publicly Owned Treatment Works (POTW) ................................................ 18 | |
| | Underground Injection .................................................................. 18 | |
| | Solid Wastes .................................................................................... 19 | |
| | Environmental Impacts of Concentrate Discharge ..................... 19 | |
| **Chapter 4** | **Environmental Issues** ................................................................. 21 | |
| | *Berrin Tansel and Conrad G. Keyes, Jr.* | |
| | Introduction ..................................................................................... 21 | |
| | Characteristics of Concentrate ...................................................... 21 | |
| | Environmental Concerns ............................................................... 22 | |
| | Discharge of Concentrate to Oceans and Bays ........................... 23 | |
| | Discharge of Concentrate to Surface Waters and Sanitary Sewers .................................................................................. 25 | |
| | Discharge by Deep Well, Land Application, and Evaporation Ponds .......................................................................... 27 | |
| **Chapter 5** | **Economic Evaluation** ................................................................. 35 | |
| | *Conrad G. Keyes, Jr. and Michael P. Fahy* | |
| | Introduction ..................................................................................... 35 | |
| | National Research Council Economic Recommendations ......... 35 | |
| | Summary of Southern California Technologies .......................... 37 | |
| | Typical Economic Summary ......................................................... 37 | |

| | | |
|---|---|---|
| **Chapter 6** | **Implementation/Case Studies** ................................................... **39** | |
| | *Conrad G. Keyes, Jr., Michael P. Fahy, and Berrin Tansel* | |
| | Introduction ................................................................................. 39 | |
| | Final Disposal Options Used ...................................................... 39 | |
| | Summary of Case Studies in Appendix A ................................... 39 | |
| **Appendix A-1** | **Oceans and Bays Discharge Case Studies** ............................ **43** | |
| | Proposed Carlsbad Seawater Desalination Plant, Carlsbad, CA .. 44 | |
| | *Nikolay Voutchkov* | |
| | Marin Municipal Water District (MMWD) Desalination Project, San Rafael, CA ........................................................ 50 | |
| | *Val S. Frenkel* | |
| | The Charles Meyer Desalination Facility, Santa Barbara, CA .... 60 | |
| | *Val S. Frenkel* | |
| **Appendix A-2** | **Sanitary Sewer and Surface Water Disposal Case Studies** ... **67** | |
| | Joe Mullins Reverse Osmosis Water Treatment Facility, Melbourne, FL ....................................................................... 68 | |
| | *Berrin Tansel* | |
| | Ormond Beach WTP Low Pressure Reverse Osmosis (LRPO) Expansion, Ormond Beach, FL ............................................ 73 | |
| | *Khalil Z. Atasi and Colin Hobbs* | |
| | Pilot-Research Membrane Treatment of Non-Irrigation Season Flows in the Rio Grande River, El Paso, TX ....................... 79 | |
| | *Michael Fahy* | |
| **Appendix A-3** | **Deep Well Injection, Land Disposal, and Evaporation Ponds** ................................................................ **89** | |
| | EPWU Deep Well Injection Plant, El Paso, TX ......................... 90 | |
| | *Michael Fahy and Kenneth Mercer* | |
| | North Collier County Regional Water Treatment Plant, Naples, FL .............................................................................. 99 | |
| | *Berrin Tansel* | |
| | Central Plantation Water Treatment Plant, Plantation, FL ........ 106 | |
| | *Berrin Tansel* | |
| | Dalby Stage 2 Desalination Plant, Queensland, Australia ........ 111 | |
| | *James Jensen* | |
| **Appendix A-4** | **Zero Liquid Discharge (ZLD) and Near ZLD** ...................... **117** | |
| | Water Desalination Concentrate Management and Piloting Study, South Florida ....................................................................... 118 | |
| | *Sandeep Sethi* | |
| | Closed-Loop Water Recovery and Recycling for Space Applications, NASA, Kennedy Space Center, FL .............. 125 | |
| | *Berrin Tansel* | |
| Index | ................................................................................................................. 131 | |

# Chapter 1 – Introduction

Conrad G. Keyes, Jr., Professor & Dept Head Emeritus, New Mexico State University, 801 Raleigh Road, Las Cruces, NM 88005, cgkeyesjr@q.com

Michael P. Fahy, El Paso Water Utilities, El Paso, TX 79961

Kenneth Mercer, AWWA, 6666 W. Quincy Avenue, Denver CO 80235

Background of EWRI work

In 2004, there were no industry-wide concentrate management performance standards for the types of desalination and water reuse technologies identified in *"Desalination and Water Purification Technology Roadmap"* [U. S. Bureau of Reclamation and Sandia National Laboratories, 2003], especially for inland desalination facilities that do not have ready access to ocean disposal. Additionally, brackish and sea-water desalination and concentrate management regulations vary significantly from state-to-state, region-to-region, and internationally, in terms of both field testing and monitoring requirements. Much of the United States contains extensive brackish ground water resources [Krieger *et al.*, 1957]. Since much of this supply underlies more easily-accessible and higher-quality fresh water resources, it has remained primarily untapped; but as fresh water resources become increasingly scarce and water demands increase, treatment of brackish water sources has gained consideration, especially as desalination technologies are improved and costs become more competitive.

In 2005, several organizations including the American Water Works Association, Ground Water Protection Council, Water Reuse Foundation, and the Environmental Protection Agency agreed to work cooperatively through an American Society of Civil Engineers (ASCE), Environmental and Water Resources Institute (EWRI) group to develop a consensus–based assessment and provide recommendations and guidance on sound and commonly acceptable concentrate management practices for new and existing desalination and water reuse facilities.

Through ASCE/EWRI, and with the support of Sandia National Laboratories, working group members were recruited from a wide-range of interested water professionals; including water resource managers, water technology developers, water utilities, and regulatory agencies; this group became the Concentrate Management Working Group (CMWG). The CMWG was designed to leverage the expertise of government, industry, and research organizations involved in desalination, water reuse, technology evaluation, and environmental protection to provide broad perspective on concentrate management issues.

The initial meeting of the CMWG (officially called the EWRI Task Committee (TC) on CM in Desal) was held in September 2004, in Phoenix, AZ in conjunction with the Water Reuse Association national conference. Subsequent meetings of the

CMWG were held in December 2004 (Las Vegas, NV) and in March 2005 (Phoenix, AZ).

During the 2007 EWRI Congress in Tampa, FL, the members of the EWRI task committee and others made presentations under the organized Desalination and Water Reuse Track generated by Sandeep Sethi. A regular funded meeting of the EWRI task committee was also conducted during the annual ASCE/EWRI Congress in Tampa, Florida on May 15, 2007. Attendees included Conrad Keyes (Chair), Ernie Avila (suggested TC Rep to the Water, Wastewater, & Stormwater Council of EWRI from the task committee), Amit Pramanik, Clayton Johnson, Findlay Edwards, Kenneth Mercer, Sandeep Sethi (Recording Secretary), and Val Frenkel. The following areas were determined to be the major viable options for concentrate management and subcommittees were formed to study each option:

>Concentrate Management to Oceans and Bays
>Discharge of Concentrate to Surface Waters and Sanitary Sewers
>Discharge by Deep Well, Land Disposal, and Evaporation Ponds
>Zero Liquid Discharge

It was subsequently decided to have each subcommittee generate case studies in their respective areas.

Case Study Topical Areas

This Committee Report (or Monograph) is a summary of Case Studies associated with concentrate management in desalination in each of the areas defined previously. The members (or authors) from the 2010-2011 EWRI task committee for the development of CM in Desal Case Studies that have been involved in providing case studies in relation to the topics of the current four subcommittees of the EWRI Concentrate Management in Desalination (CM in Desal) technical committee of FY 2009 are (see Appendix A):

>Discharge of Concentrate to Oceans & Bays – Nikolay Voutchkov and Val S. Frenkel

>Discharge of Concentrate to Sanitary Sewer and Surface Waters – Khalil Atasi & Colin Hobbs, Michael Fahy, and Berrin Tansel

>Discharge of Concentrate to Deep Well Injection, Land Applications, & Evaporation Pond – Michael Fahy, Kenneth Mercer, and Berrin Tansel

>Zero Liquid Discharge Concentrate Disposal Systems – Sandeep Sethi and Berrin Tansel

The task committee officers (Chair – Conrad Keyes, Vice Chair – Michael Fahy, and Secretary – Berrin Tansel) followed this process in the creation of this Committee

Report:

(1) Determined the appropriate case studies generated by the CM in Desal subcommittees that could be used from the four topical subcommittee areas for the document;

(2) Discussed the required introduction and general material and began the selection of additional case studies as needed at its first meeting in El Paso, TX on October 9, 2009;

(3) Selected appropriate authors and/or current material for the four major areas as designated by the subcommittee topics to be included in the document;

(4) Reviewed the developed materials at other task committee activities in FY 2010;

(5) Promoted the activities of this task committee among other professional organizations;

(6) Reviewed and combined updated materials and generated a Final Draft for review and comments by the Desalination & Water Reuse technical committee during 2011;

(7) Revised the Final Draft for the concluding review by the Desalination & Water Reuse technical committee before the end of FY 2011; and

(8) Prepared the final materials for the EWRI publications process.

Scope of the Monograph

The following chapters and appendices are provided in this Committee Report:

Chapter 2 - Overview of Processes and Configurations – Sethi & Frenkel & Mercer

Chapter 3 - Regulatory Setting – Mercer

Chapter 4 - Environmental Issues – Tansel & Keyes

Chapter 5 - Economic Evaluation – Keyes & Fahy

Chapter 6 - Implementation/Case Studies – Keyes & Fahy & Tansel

Appendix: Concentrate Management Case Studies

Appendix A-1 Ocean and Bays Disposal (Subcommittee Chair – Jim Jensen)

Nikolay Voutchkov – Carlsbad, CA plant
Val Frenkel – Marin Municipal WD plant, San Rafael, CA
Val Frenkel – Charles Meyers WW plant, Santa Barbara, CA

Appendix A-2 Sanitary Sewer or Surface Water Disposal (Subcommittee Chair – Harold Thomas)

Mike Fahy, John Balliew, & Anthony Tarquin – Pilot Research of Non-Irrigation Season
    Flows to River
Khalil Atasi and Colin Hobbs – Ormond Beach, FL (dual – land application/sewer)
Berrin Tansel - Joe Mullins RO plant

Appendix A-3 Deep well Injection, Land Disposal, and Evaporation Ponds (Subcommittee Chair – Ken Mercer)

Mike Fahy & Kenneth Mercer & Scott Reinert - EPWU Kay Bailey Hutchinson plant
Berrin Tansel - North Collier Regional plant
Berrin Tansel – Melborne, FL plant
James Jensen – Dalby Stage 2 plant, Queensland, Australia

Appendix A-4 Zero Liquid Discharge (ZLD) and Near ZLD (Subcommittee Chair – Sandeep Sethi)

Berrin Tansel - NASA Closed Loop
Sandeep Sethi – South Florida WMD

## References

Krieger, R. A., Hatchett, J. L., and Poole, J. L., (1957). "Preliminary Survey of the Saline-Water Resources of the United States", Geological Survey Paper 1374, U.S. Geologic Survey, Washington, D.C.

Treanor, P. and V. S. Frenkel (2009). "Desalination Considerations", *Civil Engineering,* ASCE, Vol. 78, No. 6, June 2009, Reston, VA, pp. 50-55.

U.S. Bureau of Reclamation and Sandia National Laboratories (2003). Desalination and Water Purification Technology Roadmap, Report for Program #95, A report of the Executive Committee, January 2003

# Chapter 2 – Overview of Desalination Processes and Configurations

Sandeep Sethi, Carollo Engineers, 401 North Cattlemen Road, Suite 306, Sarasota, Florida 34232, SSethi@carollo.com, Work: 941.371.9832, Fax: 941.371.9873

Val S. Frenkel, Kennedy/Jenks Consultants, 303 Second Street, Suite 300 South, San Francisco, CA 94107, ValFrenkel@KennedyJenks.com, Work: 415.243.2150, Fax: 415.896.0999

Kenneth Mercer, AWWA, 6666 W. Quincy Avenue, Denver CO 80235

Introduction

The by-product or residual stream generated during desalination of brackish water, seawater, or treatment of wastewater for reuse applications is termed 'concentrate'. This chapter provides an overview of desalination processes and conventional concentrate management methodologies, including citations to the configurations of the Case Studies in Appendix A.

The two major, commercialized technologies for desalination include membrane and evaporative (also known as 'thermal') technologies. While evaporative technologies have been extensively used in the Middle East, the United States has almost exclusively relied on membrane processes for desalination. Additionally, in recent years the use of membranes for desalination has become more predominant in the world in comparison to evaporative technologies due to technological advancements and cost advantages of membrane processes. The focus of this document as it relates to concentrate management is on the membrane desalination processes.

Membrane Desalination Processes

The commonly accepted definition of desalination or desalting is separation of dissolved salts and minerals from water. The sum total of dissolved constituents can be expressed as the bulk parameter Total Dissolved Solids (TDS) and measured in mg/L or ppm. TDS has a direct relationship with conductivity and a rule of thumb expression is:

TDS (ppm) = 0.67 x conductivity ($\mu$S/cm)     (Eq. 2-1)

There are three general classes of membrane desalting processes, namely:

- Reverse osmosis (RO)
- Nanofiltration (NF), and
- Electrodialysis (ED) and electrodialysis reversal (EDR).

RO and NF are pressure driven processes while ED/EDR is an electrically driven process. While all three processes are utilized in the industry, RO dominates the municipal industry in the areas of seawater and brackish water desalination. NF membranes are formulated specifically for rejection of multivalent ions, and are mostly employed for removal of hardness (i.e., $Ca^{2+}$ and $Mg^{2+}$) and/or organic carbonl. While RO is used over a wide range of TDS, ED/EDR technologies are generally used to treat lower TDS waters (typically <10,000 mg/L) or when the source water is characterized with relatively high levels of silica ($SiO_2$).

*Basic Configuration*

A general schematic of an RO desalting process is included in Figure 2.1. Water that passes through the RO membrane is termed product or 'permeate'. Permeate is actually a solution comprised of mostly water and small amounts of certain constituents (e.g. monovalent ions such as sodium and chloride) that permeate through the membrane. The remaining solution of water and constituents rejected by the membrane is termed 'reject' or 'concentrate', and this stream requires proper management and/or disposal. Constituents rejected by the membrane are concentrated in the reject stream relative to their levels in the source water, with the concentration factor a function of the system recovery.

In the case of a seawater or brackish surface water source, an additional pretreatment step is typically required to reduce the loading of suspended solids to the RO process, since a high loading of suspended solids and particulates can foul RO membranes. A case study showing this additional pretreatment step is provided in Appendix A.

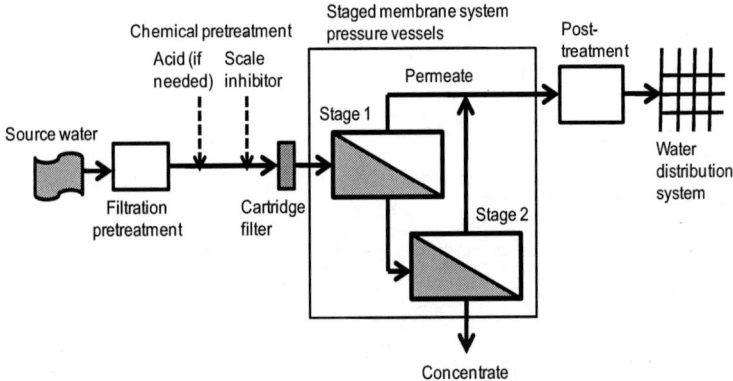

Figure 2.1 Schematic of Typical Membrane Treatment Process for Brackish Water Desalination (Courtesy of Sandeep Sethi)

## Basic Process and Design Aspects

A single membrane 'element' or 'module' has a large amount of membrane area. Spiral wound membrane elements are the most common configuration in practical use with RO systems. Hollow fine fiber RO elements are also available but are less common. Each spiral wound membrane element contains several hundred feet of membrane surface area and each element recovers about 5 to 15 percent of the feed water flow by converting it into the permeate stream.

Multiple membrane elements are typically housed in series inside a pressure vessel. A 'stage' of membranes consists of multiple pressure vessels in parallel. Most brackish water RO desalination facilities operate with two stages of pressure vessels to increase water recovery. A membrane array refers to a collection of pressure vessels over two or more stages.

The productivity of RO membranes tends to decline over time as fouling or scaling occurs (Figure 2.2). Fouling occurs when suspended solids, particulates, microbial cells, and organic matter either deposit or adsorb on the membrane surface and/or on the feed channel spacer, reducing the flow of water. Scaling occurs when dissolved solutes precipitate on the membrane surface and/or on the feed channel spacer. Fouling of the RO membranes can be grouped to the four major categories illustrated in Figure 2.2.

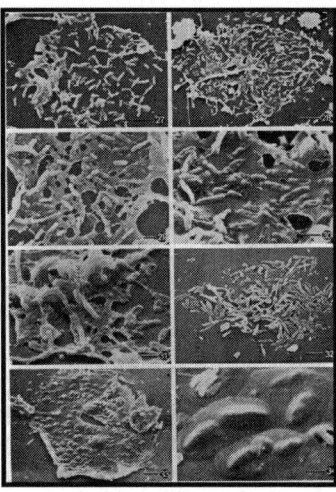

Figure 2.2 RO Membrane Fouling: Four Major Categories (Reprinted with permission from AMTA – this was originally presented as part of the AMTA/NWRI – Membrane Bioreactors Pre-Conference Workshop during the AMTA/SEDA 2008 Joint Conference, Naples, FL)

A certain level of pretreatment is typically provided upstream of the RO membranes to control fouling and scaling. Pretreatment can include pH adjustment and/or antiscalant addition (also known as scale inhibitor), followed by cartridge filtration. Cartridge filters often serve as the minimal/final safety barrier for protecting the membranes from suspended solids in the feed stream, reducing the potential for membrane fouling. While cartridge filters are typically sufficient for addressing suspended solids in the case of groundwater sources, surface waters with relatively higher levels of suspended solids often times require an additional pretreatment step based on filtration. This additional step can involve conventional treatment (i.e. coagulation/flocculation/sedimentation followed by media filters) or low-pressure membranes (i.e. microfiltration or ultrafiltration).

Characteristics of Concentrate Stream

The key characteristics of the concentrate stream are its flowrate and water quality (especially TDS). Both of these in turn impact the method and related permitting requirements for concentrate management, as well as the consequent costs.

*Recovery and Concentrate Volume*

The efficiency of conversion of feed to product water, termed recovery, depends on the source water characteristics but typically ranges between 65 to 85 percent for brackish water RO desalination and from 40 to 60 percent for seawater RO desalination. Thus, the concentrate volume ranges from 15 to 35 percent of the feed stream for brackish water RO, to as much as 40 to 60 percent of the feed stream for seawater RO (Sethi *et al.*, 2007). Disposal of the concentrate stream can result in the loss of a significant portion of the feedwater.

The levels of certain sparingly soluble salts can reach or exceed their respective saturation levels due to their high concentrations in the concentrate, particularly for brackish water desalination. Common sparingly soluble compounds of concern in RO applications include calcium carbonate, calcium sulfate, barium sulfate, strontium sulfate, and silicate. Chemicals such as antiscalant and/or acid are typically added to the feed stream to control scaling of salts that would be at or above their normal saturation levels in the concentrate. For seawater desalination, the recovery is relatively lower and typically limited by the permissible driving pressure rather than the levels of sparingly soluble salts.

*Quality Characteristics*

The level of rejected constituents in the concentrate is much greater than in the feed stream. Recovery (of water) and rejection (of pollutants) are defined as follows (Bergman, 2007):

System Recovery:  $R = \dfrac{Q_p}{Q_f}$  (Equation 2-2)

Rejection:  $r = \left(1 - \dfrac{C_p}{C_f}\right)$  (Equation 2-3)

As an example, assuming 100 percent rejection of a contaminant, its concentration level will be 4 times its feed concentration at a recovery of 75 percent (i.e., 1/(1-0.75) = 4), and 5 times its feed concentration at a recovery of 80 percent (i.e., 1/(1-0.8)=5). Section 4.2 provides further details on the water quality characteristics of the concentrate.

Types of Concentrate Management Methods

Although small concentrate streams may be disposed of by dilution in wastewater collection systems, larger concentrate streams are usually disposed of through surface water discharge to a receiving body or through deep-well injection. Land application, evaporation ponds, and thermal evaporation are other, relatively less common methods of concentrate disposal (Frenkel, 2004 & 2009). The types of concentrate disposal methods are listed below and discussed in more detail in the following section, and case studies of each are provided in Appendix A.

- Surface Water Discharge
    - Ocean and bays
    - Desalination plant co-location
    - Rivers, canals, and lakes
- Deep-well injection
- Sanitary sewer
- Land disposal
- Evaporation ponds
- Thermal evaporation

In general, although a range of disposal options are available, there are significant challenges associated with the permitting (due to environmental impacts) and/or cost of each option. These challenges are critical and can limit the feasibility of a desalination project, especially for inland communities. The environmental impacts and mitigation measures for these concentrate disposal methods are discussed in Chapter 4 and some economics associated with a case study is included in Chapter 5.

*Concentrate Minimization*

In the last decade, there has been increased emphasis on investigating and developing methods for concentrate minimization. This can be achieved via additional treatment of concentrate via existing unit processes, or potentially through emerging desalination processes such as forward osmosis (FO) or membrane distillation (MD), amongst others (Sethi et al., 2009). For example, additional treatment can include chemical softening of the concentrate stream, followed by re-processing through another membrane desalination process. Conventional approaches such as thermal evaporation can also minimize the volume of concentrate requiring ultimate disposal. However, as discussed in Section 2.5, the cost of thermal evaporation processes is currently prohibitive for concentrate treatment in typical municipal desalination plants. In general, these thermal evaporation processes can be used to achieve either zero-liquid-discharge (ZLD) or near-ZLD in areas where other conventional disposal methods are not feasible due to lack of access and/or other considerations.

Concentrate Management and Disposal Methods

As discussed in Section 2.4, the conventional methods for concentrate management include surface water discharge, sewer discharge, deep well injection, evaporation ponds, land application, and thermal evaporation towards zero liquid discharge or near-zero liquid discharge applications.

*Surface Water Discharge*

Disposal of concentrate through surface water discharge to a receiving body can be practiced for both seawater and inland desalination. This is the most common concentrate disposal practice in the United States and is employed by approximately 41 percent of all desalting facilities in the US (Mickley, 2006). A NPDES (National Pollutant Discharge Elimination System) permit is required for concentrate discharge to surface water.

1. Ocean and Bays Disposal (Frenkel et al., 2010)

Seawater desalination facilities tend to be large in size and have a low process recovery resulting in large concentrate volumes. These facilities are typically sited by the ocean and the large concentrate volume is disposed of back to the nearby ocean via an outfall structure. The outfall essentially constitutes a long pipeline to a certain depth and distance into the ocean, with a diffuser structure at the end. The outfall pipeline and diffuser are designed by modeling the plume to provide the dilution/dispersion levels and mixing zones required by the permit, such that the salinity levels and marine life in the receiving body are not adversely affected. Independent seawater desalination plants, i.e. those that are not collocated with a power plant require a new outfall structure for concentrate disposal.

Brackish water desalination plants can also discharge to ocean and bays if the conveyance distance is not prohibitive, although sometimes a long distance 'brine line' can be employed
In such cases, the concentrate is typically blended with the wastewater from a

municipal wastewater treatment facility, and the existing outfall of the wastewater facility is utilized for the combined disposal. An example is the Santa Ana Regional Interceptor (SARI) which is a 93-mile long gravity pipeline in California (USBR, 2009). While conveying concentrate to such long distances is economically prohibitive for any single inland desalination facility, it becomes a viable option when the cost and fees are shared by multiple facilities discharging into the common brine line. The same line can receive treated wastewater discharged from the wastewater treatment plants reducing combined salinity of the effluent at the same time.

2. Desalination Plant Co-location

Seawater desalination plants are typically collocated with a power plant to gain the advantages of the existing intake and outfall infrastructure. Thus, in this case, the existing outfall structure at the power plant is utilized for the concentrate disposal. A portion of the warmer return cooling water from the power plant is used for desalination as the higher temperature provides a slightly higher permeate flux across the membrane. The concentrate is blended and diluted with the remaining return cooling water and returned to the ocean (Frenkel *et al.*, 2010).

3. Discharge to Rivers, Canals, and Lakes

Inland brackish water plants can also discharge concentrate to an accessible surface water body such as a river or canal. As in the case of ocean disposal, a NPDES permit is required and the permit limits may include total dissolved solids (TDS), total suspended solids (TSS), and specific nutrients and metals (e.g. arsenic). This method is economically viable if the length of the pipeline to the receiving surface water body is reasonable and the concentrate meets the permit requirements. Nonetheless, due to the potential environmental impacts on marine life, the permitting considerations are usually challenging.

*Sewer Discharge*

With appropriate approvals or permit from the local sewage agency, the concentrate from inland desalination facility can be discharged to an existing sewage collection system. This is the second most common concentrate disposal practice in the United States and is employed by approximately 31 percent of all desalting facilities in the United States (Mickley, 2006). Basic pretreatment such as pH neutralization is usually required and other requirements and limitations can also be imposed. These are put in place to protect the sewers and wastewater treatment plants infrastructure and treatment process, as well as the final wastewater effluent and biosolids quality. This disposal method is typically feasible and economical only for relatively smaller discharge volumes that have limited permitting requirements. Also see the discussion on brine lines above for a regional concept of a concentrate collection system.

### Deep Well Injection

Subsurface discharge of deep well injection (DWI) disposes the concentrate into a deep geological formation, which will serve to permanently isolate the concentrate from shallower aquifers that may be used as a drinking water source. Approximately 17 percent of desalting facilities use this method for concentrate disposal (Mickley, 2006). Regulatory considerations include the receiving aquifer's transmitivity and TDS, and the presence of a structurally isolating and confining layer between the receiving aquifer and any overlying source of drinking water.

DWI injection is typically economical and employed only for larger concentrate flows (> 1 mgd) and thus used for larger RO plants (Malmrose *et al.*, 2004).

### Evaporation Ponds

Disposal of concentrate via an evaporation pond is essentially based on the natural phenomenon of atmospheric evaporation. In this method the concentrate is pumped into a shallow lined pond and allowed to evaporate naturally using solar energy. Once the water has evaporated, the salt sludge is either left in place or removed and hauled offsite for disposal. Evaporation ponds can be a viable option in relatively warm, dry climates with high evaporation rates, level terrain, and low land costs.

This disposal method can be expensive due to the large surface area required and the associated land and impermeable liner costs (NRC, 2004). They are typically economical and employed only for smaller concentrate flows. Regulatory requirements, ecological impacts, and possible concentration of trace elements to toxic levels may determine the design, construction, and operation of evaporation ponds (ASCE, 1990).

A solar energy pond is a special type of evaporation pond that focuses on capturing solar energy with the goal to use it beneficially; typically in the overall desalination process to mitigate the process energy needs. The approach uses salinity gradients to trap energy in the lower, higher density layer of the concentrate in the pond. The solar energy penetrates the upper, less concentrated layers. The lower, heated layer does not rise, due to the higher concentration and density and the absence of convection, and thus reaches significantly high temperatures. The energy thus trapped in this layer is extracted through the use of heat exchangers.

### Land Application

Land application such as spray irrigation is a beneficial reuse of concentrate. It can be used for lawns, parks, golf courses, or crop land. Land application depends on the availability and cost of land, percolation rates, irrigation needs, water quality tolerance of target vegetation to salinity, and the ability to meet ground water quality standards.

### Thermal Evaporation towards Zero Liquid Discharge or Near-Zero Liquid Discharge

Thermal evaporation processes (such as brine concentration via vapor compression, followed by brine crystallization via crystallizer or spray dryer), or their combinations, can be used to achieve zero-liquid-discharge (ZLD), i.e., reduction of the concentrate to essentially dry solids such that no liquid is discharged from the site. The solid product is typically disposed via landfill disposal. Near ZLD processes (e.g. brine concentration via vapor compression) do not include a crystallization step and reduce to the concentrate to slurry rather than all the way to a solid product.

While these methods are well established and developed, both are characterized with high capital costs and are not feasible for large concentrate flows. Additionally, thermal evaporation processes are energy intensive and typically not considered cost effective for municipal water treatment, especially for large applications. Thus, capital and operating costs of these methods is currently very high and can exceed the cost of the desalting facility. Therefore, the ZLD option is typically not employed except for special situations (e.g. inland desalination facility with no sewer or surface water access) coupled with very small concentrate flows.

#### Pilot Testing of Concentrate Minimization Methods

Conventional concentrate disposal methods are well established and typically not pilot tested. However, the main desalting process such as brackish or seawater RO is usually pilot tested for process validation for site specific conditions, optimizing pretreatment, refining design criteria, and/or operator training etc.

Concentrate minimization via non-thermal technologies and approaches is a developing area that is active with pilot testing. As discussed in Section 2.4, this can be achieved via additional treatment of concentrate via existing unit processes, such as chemical softening followed by re-processing through another membrane desalination process such as RO or EDR. Other processes such as biological reduction have also been tested (Sethi *et al*, 2009). Pilot and bench-scale testing these processes has been performed at several sites in Florida, California, and Texas (Carollo Engineers, 2009).

Although the unit process involving such chemical softening are themselves well established, bench or pilot testing may be required to optimize the operating and design conditions for treatment of the site-specific concentrate stream. Because the raw water quality and recovery of the primary desalination varies at each facility, this results in different concentrate water quality that requires treatment. The different concentrate quality, e.g. different hardness levels and/or different sparingly soluble salts, requires specific testing for optimizing chemical doses; as well as evaluating and optimizing the parameters for the secondary RO or EDR process.

Data collected and/or analyzed during such pilot testing comprises the water quality of the primary RO feed, primary concentrate, and water quality at each treatment step of the primary concentrate (e.g. post softening, post filtration, and post secondary RO permeate and final concentrate). The data required for evaluating the hydraulics performance of each concentrate treatment step is also collected and analyzed (e.g. flows, pressures, etc.).

References

ASCE (1990). *Agricultural Salinity Assessment and Management* (Manual No. 71), American Society of Civil Engineers, Reston, VA.

Bergman, R. (2007). *Reverse Osmosis and Nanofiltration (M46)*, Second Edition Ed., American Water Works Association, Denver, CO.

Carollo Engineers (2009). "Water Desalination Concentrate Management and Piloting." South Florida Water Management District (SFWMD), December 2009.

Frenkel, V. (2004). *Using Membranes to Manage Salinity*. 2004 National Salinity Management and Desalination Summit, Concentrating on Solution, December 13-14, 2004, Las Vegas, Nevada.

Frenkel, V. (2008). *Membrane Technology Fundamentals and Fouling Control*. Pre-Conference Workshop AMTA/NWRI – Membrane Bioreactors (MBRs) at the American Membrane Technology Association and SEDA Joint Conference, AMTA/SEDA-2008, 14-17 July, 2008. Naples, FL.

Frenkel, V. (2009). *Salinity & Brine Management Road Map*. Multi-State Salinity Coalition, MSSC 2009 Water Supply, Agriculture & Salinity Management Workshop, September 29-30, 2009, Indian Wells, California.

Frenkel, V., and Treanor, P. (2010). *Bay vs. Ocean Desalination Concentrate*. American Membrane Technology Annual Conference, AMTA-2010, July 12-15, 2010, San Diego, California.

Malmrose, P., Lozier J., Mickley, M., Reiss, R., Russell J., Schaefer, J., Sethi, S., Manuszak, J., Bergman, R., and Atasi, K.Z. (2004). Committee Report: Current Perspectives on Residuals Management for Desalting Membranes, *Journal of the American Water Works Association*, Volume 96, No.12. December 2004.

Mickley, M.C. (2006) *Membrane concentrate disposal: practices and regulation*. Desalination and Water Purification Research and Development Program Report No. 123 (Second Edition). U.S. Department of Interior, Bureau of Reclamation, April 2006.

NRC (2004). "Review of the desalination and water purification technology roadmap" The National Academic Press, Washington, D.C.

NRC (2008). "Desalination: A National Perspective" The National Academic Press, Washington, D.C.

Sethi, S., Xu, P. and Drewes, J. E. (2007). "New Desalination Configurations and Technologies for Recovery Increase and Concentrate Minimization." Proceedings of the World Environmental and Water Resources Congress 2007, ASCE and Environment and Water Resources Institute, Tampa, FL, May 15-19 2007.

Sethi, S., Walker, S., Xu., P., and Drewes, J. (2009). "Desalination Product Water Recovery and Concentrate Volume Minimization." Water Research Foundation, *Report Number 91240*, Denver, CO, 2009.

Sethi, S., MacNevin, D., Munce, L., Akpoji, A., Elsner, M., and An, J.H. (2010). "Results from Concentrate Minimization Study for Inland Desalination in South Florida." Proceedings of the American Water Works Association Annual Conference & Exposition, Chicago, IL, June 20 - 24, 2010.

USBR (2009). *Southern California Regional Brine-Concentrate Management Study – Phase I Institutional Issues.* U.S. Department of the Interior, Bureau of Reclamation.

# Chapter 3 - Regulation of Concentrate Management

Kenneth Mercer, AWWA, 6666 W. Quincy Avenue, Denver CO 80235

Concentrate Management Regulation Overview

Membrane concentrate and spent cleaning solutions are classified as industrial wastes in the United States by federal regulations. Discharge and assimilation of concentrate into the environment is often the most critical regulatory issue when implementing a desalting process (Water Desalting Committee of the American Water Works Association, 2004). Regulations governing residual streams disposal or reuse account for environmental and toxicological impacts by evaluating the nature of the potential receiving body and the quality and flow of the waste stream; leaving federal, state, and regional agencies to lay the framework for the requirements that must be met before residuals streams disposal or reuse is allowed. Each state has its own often unique scheme for regulating desalination residual streams (AWWARF, 2000; USBR, 2009); however, uniform requirements are generally established for all states under several statutes promulgated by the following United States federal regulatory agencies:

U.S. Environmental Protection Agency (USEPA)
U.S. Army Corps of Engineers (USACE)
U.S. Fish and Wildlife Service (USFWS)
NOAA Fisheries Service, a division of the U.S. Department of Commerce

The bulk of concentrate management regulations are promulgated by USEPA as described in the following areas.

National Pollutant Discharge Elimination System

Perhaps the most far-reaching regulatory provisions concerned with residuals streams disposal have been established by US EPA; which, under the Clean Water Act of 1972, created the National Pollutant Discharge Elimination System (NPDES) permit program. The NPDES permit program regulates point source discharges of industrial and municipal facilities (called *direct discharger*) into surface waters and wetlands, including drinking water treatment plants (DWTPs). The NPDES program was established to reduce or eliminate water pollution that can make surface waters unsuitable for drinking, fishing, swimming, and other activities.

A DWTP that is a direct discharger must hold a NPDES permit and may only discharge pollutants in conformance with the terms of that permit, i.e., the DWTP is responsible for treating residual streams to the levels prescribed in the discharge permit (US EPA, 2005). Discharge limits, as defined in Total Maximum Daily Loads (TMDLs), may restrict total suspended solids (TSS), total dissolved solids

(TDS), salinity, or specific pollutants such as nutrients, arsenic, or barium (Bergman, 2007). In addition to effluent limitations, NPDES permits typically impose various requirements involving operation and maintenance, monitoring, reporting, and record keeping (Pontius *et al.*, 1996).

Residuals Treatment Guidelines for Discharge to Publicly Owned Treatment Works (POTW)

US EPA Title 40 parts 122 and 403 detail how removal of pollutants can help facilities meet pretreatment standards for discharges to POTW or meet the effluent limits of an NPDES permit. Title 40 subchapter N "Effluent Guidelines and Standards" applies to indirect discharge of residuals, requiring a reduction of the volume and toxicity of hazardous wastes to the degree it is economically practical. This USEPA regulation is intended to fulfill three objectives ("General Pretreatment Regulations for Existing and New Sources of Pollution."):

> To prevent the introduction of pollutants into POTW that will interfere with its operation, including interference with use or disposal of municipal sludge;
> To prevent the introduction of pollutants into POTW which will pass through the treatment works or otherwise be incompatible with such works; and
> To improve opportunities for recycle of reclaim municipal and industrial wastewaters and sludges.

Overall, Title 40 establishes the responsibilities of federal, state, and local governments, industry, and the public to control pollutants (concentration and flow) that pass through or interfere with treatment processes in POTW or which may contaminate sewage sludge. In addition, direct dischargers are subject to local limits, and if a pollutant has both a local and federal limit, the more stringent of the two is normally applied.

Underground Injection

Discharges to groundwater are also regulated by US EPA through the underground injection control (UIC) regulations under the Safe Drinking Water Act (SDWA). States with requirements at least as stringent as those of US EPA often administer the UIC program. Underground injection is the technology of placing fluids underground, in porous formations of rocks, through wells or other similar conveyance systems (US EPA, 2005). The SDWA established the UIC Program to ensure that injection wells do not endanger current and future underground sources of drinking water (USDW). The UIC Program groups underground injection into five classes for regulatory control purposes, where the well class defines the quality of waste that may be injected. Discharge of sludge, brine, or other fluids are prohibited if they will cause any USDW to exceed any SDWA maximum contaminant level or otherwise affect public health (Pontius, *et al.*, 1996).

Solid Wastes

The Resource Conservation and Recovery Act, enacted by Congress in 1976, established a system for managing non-hazardous and hazardous solid wastes in an environmentally sound manner. As they may be hazardous, all solid wastes generated from concentrates or brines (such as residues or salt/mineral mixtures following evaporation) would have to be managed from the point of origin to the point of final disposal. Waste that is generated should be treated, stored, or disposed of so as to minimize the present and future threat to human health and the environment. It should be noted that membrane concentrates are generally considered a solid waste (Pontius *et al.*, 1996).

Environmental Impacts of Concentrate Discharge

Another important set of regulations concerned with residuals disposal is the National Environmental Policy Act (NEPA) (US EPA, 2005), which requires federal agencies to integrate environmental values into their decision making processes by considering the environmental impacts of their proposed actions and reasonable alternatives to those actions. In many cases, an Environmental Impact Statement (EIS) must be prepared for federal and regional EPAs for review and comment before process implementation can occur. Oftentimes state agencies will take the lead for preparing the EIS, requiring project proponents to fully address such issues as potential impacts to estuarine or marine habitats. At a minimum, an environmental assessment should include 1) analysis of the source impacts, 2) analysis of the impacted ecosystem, 3) definition of the links between source and targets, 4) recommendations for mitigation measures and 5) sustainability of the environmental protection measures (Hoepner, 1999).
Permits for disposal options other than release to POTW can depend on factors that influence the regional regulatory environment, especially risks to endangered ecological systems and public perception. In addition, discharge limits for certain receiving bodies may be based on residual water quality parameters such as hardness. The state of Florida provides some good examples of how local regions can respond to the challenges posed by residuals disposal. For instance, Florida's Department of Environmental Protection (DEP) has established protocols for conducting tests to determine the major seawater imbalance toxicity of desalting membrane residuals (Florida DEP Bureau of Laboratories, 2004). In addition, Florida DEP also regulates deep-well injection through their Underground Injection Control group. These are but two examples of myriad controls which may affect implementation of desalting membrane processes.

## References

AWWARF (2000). *Current Management of Membrane Plant Concentrate.*

Bergman, R. (2007). *Reverse Osmosis and Nanofiltration (M46),* Second Edition Ed., American Water Works Association, Denver, CO.

Florida DEP Bureau of Laboratories (2004). "Technical Guidance Document for Conducting the Florida Department of Environmental Protection's Protocols for Determining Major-Seawater-Ion Imbalance Toxicity (MSIIT) in Membrane-Technology Water-Treatment Concentrate."

Hoepner, T. (1999). "A Procedure for Environmental Impact Assessments (EIA) for Seawater Desalination Plants." *Desalination,* 124(1-3), 1-12

Mercer, K.L. (2009). "Chemical treatment of high pressure membrane concentrate for improved residuals management", PhD Thesis, University of Massachusetts, Amherst, MA. http://scholarworks.umass.edu/dissertations/AAI3336976

Pontius, F. W., Kawczynski, E. and Koorse, S. J. (1996). "Regulations Governing Membrane Concentrate Disposal." *Journal American Water Works Association,* 88(5), 44-52.

Pontius, F. W. (1997). "Regulating Filter Backwash Water." *Journal American Water Works Association,* 89(8), 14-&(17 ??).

USBR (2009). Regulatory Issues and Trends Report: Southern California Regional Brine-Concentrate Management Study – Phase I Lower Colorado Region, Denver?

US EPA. (2005). "National environmental policy act (NEPA)." <http://www.epa.gov/compliance/nepa/> (April, 2005).

Water Desalting Committee of the American Water Works Association. (2004). *Water Desalting Planning Guide for Water Utilities,* John Wiley and Sons, Hoboken, NJ.

# Chapter 4 - Environmental Issues

Berrin Tansel, Professor, Florida International University, Civil and Environmental Engineering Department, Miami, FL 33174, tanselb@fiu.edu

Conrad G. Keyes, Jr., Emeritus Prof & DH; Civil, Agricultural, & Geological Engineering, NMSU, Las Cruces, NM 88005, cgkeyesjr@q.com

Introduction

Concentrate water treatment plants must be managed in an environmentally compatible manner and requires significant environmental assessment efforts depending on the quality of the concentrate and geographical location of the facility (Younus, 2005; Squire *et al.*, 1997; Mercer, 2009). Many disposal options require chemical and biological monitoring of the receiving environment (Squire *et al.*, 1997; Tularam and Ilahee, 2007) as well as mixing studies, environmental and ecological health risk assessments, and environmental quality modeling. This chapter reviews the environmental issues relevant to each of the following concentrate management options (see Appendix A for the case studies used in this document):

1. Discharge of Concentrate to Oceans and Bays (Appendix A-1)
2. Discharge of Concentrate to Sanitary Sewers and Surface Waters (Appendix A-2)
3. Discharge by Deep Well, Land Application, and Evaporation Ponds (Appendix A-3)
4. Zero Liquid Discharge or Near Zero Liquid Discharge ((Appendix A-4)

There are advantages and disadvantages to each approach. Both environmental and cost analyses should be performed to weigh the potential impacts.

Characteristics of Concentrate

Characteristics of the concentrate depends on feed water quality, membrane type, pretreatment processes used, chemicals added (e.g., antiscalants, acid, and chlorine), process configuration (recovery), and operational constraints. Depending on the membrane type and pretreatment requirements, the concentrate contains salts, dissolved organics, and microorganisms (viruses, bacteria, protozoa) and other particulates. Concentrate usually contains low concentrations of particles, typically <10 mg/L total suspended solids (Malmrose *et al.*, 2004). The concentrations of dissolved salts are typically 4 to 10 times those in the feed water for brackish source water, and 1.5 to 2.5 times the source water levels for seawater (Younus, 2005; Bergman, 2007). Critical concentrate parameters include total dissolved solids (TDS), temperature, dissolved oxygen (DO), and specific weight (density). The concentrate may also contain low levels of chemicals used during pretreatment and post-treatment processes. The chemicals used for membrane cleaning are generally targeted to remove a specific form of fouling. For example, citric acid is commonly used to dissolve inorganic scaling, and other acids may be used for this purpose as well. Strong bases such as caustic are typically employed to dissolve organic material (US EPA, 2003). Table 4.1 presents typical chemicals for membrane cleaning.

**Table 4.1 Chemical used to reduce membrane fouling**

| Chemical type | Chemical | Control Target |
|---|---|---|
| Acid | Citric acid<br>Hydrochloric acid<br>Sulfuric acid | Inorganic scale |
| Base | Caustic | Organics |
| Disinfectant | Sodium hypochlorite<br>Chlorine gas<br>Hydrogen peroxide | Biofilm |
| Surfactant | Various | Organics<br>Inorganics<br>Inert |

Source: Tansel and Sosnikhina (2009). Reproduced with permission.

Environmental Concerns

There are location and process factors associated with each concentrate management option. Therefore, no single option is ideal or most appropriate for every application. In many cases surface water discharge may be an economically feasible option if the facility is close to a surface water body. However, the permitting process may be difficult, since there are potential environmental impacts if the salinity of the concentrate is significantly higher than that of the receiving body (see Chapter 3 for more details). Ecotoxicity tests or plume mixing modeling are often required (e.g., Appendix A-1, Marin Municipal Water District Desalination Project; Appendix A-2, Joe Mullins Reverse Osmosis Water Treatment Facility, Melbourne, Florida). Discharge to the sanitary sewer may have similar issues, since the wastewater treatment process does not typically affect dissolved solids concentrations, and the treatment plant effluent may ultimately be discharged to surface water receiving body. For the land application practices, concentrate applied to soils may affect surface or ground water resources. Spray irrigation may be implemented if there is a need for irrigation close to the desalination plant and if the concentrate of dissolved solids is acceptable for crop growth. State agencies often take the lead for preparing the EIS and require the environmental issues to be fully

addressed (i.e., potential impacts to estuarine or marine habitats). At a minimum, an assessment should include (Hoepner, 1999; Younus, 2005):
1. Analysis of the source impacts,
2. Analysis of the impacted ecosystem,
3. Definition of the links between source and targets,
4. Recommendations for mitigation measures, and
5. Sustainability of the environmental protection measures.

Permits for disposal options other than release to publicly owned treatment works (POTW) can depend on factors that influence the regional regulatory environment, especially risks to endangered ecological systems and public perception. In addition, discharge limits for the receiving bodies may be based on residual water quality parameters such as hardness. For instance, Florida's Department of Environmental Protection (DEP) has established protocols for conducting tests to determine the major seawater imbalance toxicity of desalting membrane residuals (Florida DEP Bureau of Laboratories, 2004). In addition, Florida DEP regulates deep-well injection through their Underground Injection Control group. These are two examples of controls which may affect implementation of desalting membrane processes (Appendix A-2, Joe Mullins Reverse Osmosis Water Treatment Facility, Melbourne, Florida). Blending the filtered water with raw water is an effective way to provide alkalinity and pH control, thus reducing the amount of post-treatment chemicals.

Discharge of Concentrate to Oceans and Bays

Discharge of concentrate to a surface water body (i.e., oceans and bays) is the most common practice for concentrate management, primarily due to its economic advantage when water treatment plants are located in close proximity to oceans and bays and pipeline conveyance distances are not excessively long. The environmental concerns with discharge of concentrate to oceans and bays include compatibility of the concentrate with the receiving water in terms of (US EPA, 2003):

- Salinity,
- Dissolved oxygen (DO) levels,
- Dissolved gasses,
- Free chlorine,
- Alkalinity, and
- pH.

Concentrate can impact receiving environments through a combination of physical (e.g. temperature, turbidity) and chemical (e.g. trace pollutants, salinity, pH, and dissolved oxygen levels) factors which can stress marine ecosystems, altering the distribution and life cycles of existing aquatic plants and animal communities in the vicinity of discharge structures (El Fadel and Alameddine, 2005). For example, polyphosphate scale inhibitors can potentially stimulate algae blooms, thereby depleting dissolved oxygen levels; alternatively, trace pollutants such as arsenic can bioaccumulate in clams and seaweed in contaminated sites (Koch et al., 2007). A study by Abdul-Wahab et al., (2007) found that design of ocean outfalls to maximize dispersion improves water quality in the vicinity of discharge. Another study that focused on modeling optimal discharge conditions reported that adverse impacts generally occur within 300 m of the discharge point for well designed outfalls but can reach as far as 2000 m under some conditions; multiport subsurface injectors were found to provide greater dilution than surface discharge outfalls (Alameddine et al., 2007). Some facilities dilute the concentrate with surface water or groundwater, effluent from wastewater treatment plants, or cooling water so that salinity of the concentrate is adjusted so that salinity of the concentrate is not significantly higher than that of receiving water (i.e., less than 10% difference). Prior to discharge, the DO level of the concentrate must be adjusted to minimize any potential impacts on the aquatic species in the receiving stream. If the concentrate contains dissolved gasses (i.e., hydrogen sulfide, carbon dioxide), then they should be removed by aeration or other methods (Hoepner, 2002). Free chlorine, if present, must be neutralized.

In general, the concentrate sinks to the ocean bottom, establishing a boundary where the salinity exceeds the background levels. Studies at the small desalination plants in Florida, which dispose directly into the sea or use a short discharge pipe, showed no environmental impact on the animal and plant life near the outlets. Some considerations for discharge are listed below (Mickley, 2001):

- For all surface water discharges, diffusers will probably be required. Occasionally, a pipeline discharging into a high-ocean energy zone may be required for adequate dispersion of the concentrate.
- Corrosion byproducts, higher temperatures, or low oxygen concentrations can harm fresh and marine water biota. Effluent from a distillation plant may contain copper from corrosion.
- Toxicity of the effluent stream can possibly be reduced to acceptable levels by dilution with the receiving water. Some dilution is gained from power plant cooling water in a dual-purpose or co-located plant.
- Outfalls to the ocean should be located on the open coast. Locations on estuaries and areas with restricted interchange of water should be avoided.
- Concentrate from brackish water facilities may require additional treatment prior to discharge, such as aeration or pH adjustment. Mixing zones in the receiving water may be required.

Appendix A-1 presents case studies of plants which discharge to ocean and bays. In highly populated areas, disposal to oceans or bays may be a problem due to potential interference of the mixing zone with recreational areas (e.g., Appendix A-1, Marin Municipal Water District Desalination Project). Disposing of concentrate underwater through outfalls that stretch far into the ocean requires studies of the mixing zones.

Environmental regulatory agencies can establish 'allocated impact zones' within which the water quality limits may be exceeded for the specified pollutants (Appendix A-1, The Charles Meyers Desalination Facility). For the submerged ocean outfalls, an initial dilution zone is established where the mixing zone is the distance the plume travels before it contacts the ocean bottom (Kimes, 1995).

The increase in salt concentration could disturb the ecosystem. Most at risk organisms include the benthic marine organisms. Studies have shown that long abdomen invertebrates are more sensitive to high salinities than short abdomen invertebrates. Ocean conditions such as waves, tides, currents, and water depth can affect natural dilution and mixing characteristics at the concentrate disposal location (Mickley, 2001).

Dispersion of buoyant discharges should be analyzed to predict the environmental effects of concentrate disposal and dispersion rates. Factors to consider for jet dilution include density differences between the concentrate and the receiving water, and the characteristics of the jet stream (i.e., momentum and velocity) at the outlet.

Discharge of Concentrate to Surface Waters and Sanitary Sewers

*Discharge to Surface Waters*

Disposal to other surface waters includes tidal lakes, brackish canals, (Bergman, 2007). When concentrate enters the receiving water, it creates a high salinity plume in the receiving water. Depending on the density of the concentrate in comparison to the seawater, the plume sinks, floats, or stabilizes in the water. The type of dispersion and natural dilution of the concentrate plume depends on the discharge location. Without proper dilution, the plume may extend for hundreds of meters, beyond the mixing zone, harming the ecosystem along the way. Mixing zones are quantified limits within the receiving waters where the law allows surface water to exceed water quality standards due to the existence of point source disposal. State governments determine these limits and utilities monitor them. For example, Florida's mixing zone limitations are 800 meters (2,625 ft) canals, rivers, and streams; 31 acres for lakes, estuaries, bays, lagoons, and bayous; and 124 acres for oceans (Truesdall *et al.*, 1995). If the concentrate does not pass the whole effluent toxicity (WET) test and natural dilution is not sufficient to diffuse the concentrate, then desalination plants use artificial dilution methods. WET tests are required in discharge permits to measure the potential toxicity of concentrate to surface water bodies (i.e., rivers, lakes, bays, and ocean). WET testing involves biomonitoring or and/or bioassay testing. The concentrate can be diluted through efficient blending, diffusers, or within mixing zones prior to surface disposal. Blending is accomplished by mixing the concentrate with cooling water, feedwater, or other low TDS waters before disposal.

The long-term effects of concentrate disposal in the oceans are not well understood (California Coastal Commission Report 2004) and questions have arisen about the environmental impacts of these discharges (Malaxos and Morin, 1990; Tularam and Ilahee, 2007). For example, Tularam and Ilahee (2007) reported that the high density of the discharge reaches the bottom layers of receiving waters and may affect marine life particularly at the bottom layers or boundaries that receive desalting membrane concentrate (e.g., shrimp). Gacia *et al.* (2007) reported some of the negative effects of concentrate discharge on meadows of endemic seagrass Posidonia oceanica. To avoid recirculation of plant effluents to the intakes of the desalination facility, outlets are specifically engineered to discharge in coastal areas where maximum circulation patterns and hydrographic currents can easily disperse and dilute the brine (Ahmed et al., 2000).

Appendix A-2 presents case studies of plants that discharge concentrate to surface waters. Extensive mathematical modeling may be needed to simulate the hydrodynamics and particulate transport within the estuarine system as in the case of Joe Mullins Reverse Osmosis Water Treatment Facility, Melbourne, Florida (Appendix A-2). In general, direct discharge without treatment into a river, lake, or other watercourse cannot be made without degrading surface water quality. Water quality control laws of most political bodies prohibit such discharge. Because of the extensive and costly permit reviews, some plants have avoided surface water discharge in favor of other options (Skehan and Kwiatkowski, 2000). The effluent from a desalting plant located near a coast would probably be discharged into the ocean or large estuaries. An example of a facility that discharges concentrate directly to surface water is a reverse osmosis plant in the city of Newport News, Virginia, which began operation in 1998. This facility was designed to produce 21,600 cubic meters per day ($m^3/d$) of potable water from a brackish ground water supply, which has a feed water total dissolved solids (TDS) value of 2,900 mg/l and fluoride contamination. The membrane concentrate is discharged directly into a nearby river. In Florida facilities that discharge the concentrate to the ocean are Marco Island Reverse Osmosis (RO) Water Treatment Plant and Sanibel RO Water Treatment Plant.

*Discharge to Sanitary Sewers*

A National Pollution Discharge Eliminations System (NPDES) permit is not required for discharge to a POTW, however, the impact of both the concentration and discharge flow of the concentrate must meet the approval of the POTW because both will impact its NPDES permit. The high volume of some concentrates prohibits discharge to POTW, while in other cases concerns have been raised over the increased TDS level of the POTW effluent that results from concentrate addition (Mickey, 2001). Typically, a POTW's biological treatment process is affected by salinities that cause the overall wastewater TDS concentrations to exceed 3,000 mg/L (Bergman, 2007). One way to control the concentration of pollutants entering a POTW is to require pretreatment by the individual dischargers prior to release (as proposed in this study), where pretreatment could remove key pollutants to an acceptable level. Other considerations for concentrate discharge to a sewer include the distance between the two facilities, whether the two facilities are owned by the same entity, and any anticipated future capacity increases (Mickley, 2001).

The sewer capacity and wastewater treatment plant capacity should be addressed to ensure that wastewater collection can accommodate the additional flow due to concentrate discharge. The concentrate can affect the wastewater effluent quality. Potential impacts on the wastewater treatment plant effluent should be evaluated to ensure that the wastewater treatment facility is able to comply with their permit NPDES requirements. If the concentrate salinity and flow levels are significant, the operational performance of biological processes (i.e., activated sludge, membrane bioreactor) efficiency of the may be impacted. Another concern with this disposal method is the potential for TDS increase in the processed water (wastewater treatment plant effluent) and the probable reduction of plant treatment capacity. The high TDS content of treated wastewater poses an environmental concern if the plant returns the treated water into surface water systems. Hence, discharge to sewer is typically used by smaller and medium sized plants as larger plants may impact the operation of the wastewater treatment plants.

## Discharge by Deep Well, Land Application, and Evaporation Ponds
### *Deep Well Injection*

Deep well injection involves injecting concentrate into aquifers that are not used for drinking water. Depths of injection wells range from 322 meter (0.2 miles) to 2575 meters (1.6 miles) below the surface (Tsiourtis, 2001). A tubing and packer design is commonly required for monitoring of well integrity. One or more monitoring wells in proximity to the disposal well are also typically required to detect any changes to groundwater quality. Deep injection wells should also be subjected to tests for strength under pressure and checked for leaks that could contaminate adjacent aquifers (Ahmed *et al.*, 2000). High iron concentrations may result in fouling when conditions alter the valence state and convert soluble to insoluble species. Organic carbon may serve as an energy source for indigenous or injected bacteria resulting in rapid population growth and subsequent fouling. Waste streams containing organic contaminants above their solubility limits may require pretreatment before injection into a well.

Deep-well injection enables liquid wastes to be pumped into porous subsurface rock formations. Well depths can vary from 50 meters to over one thousand meters depending on the geological conditions at the site (Mercer, 2009). Because of concerns about aquifer contamination, injection wells are suitable for locations where groundwater is used for domestic or agricultural purposes, areas vulnerable to earthquakes, or regions with mineral resources (Ahmed et al. 2000). Monitoring wells are typically installed to ensure the integrity of the boundaries. Injection zones must have a TDS level greater than 10,000 mg/L and at least one overlaying, confining layer (Bergman, 2007).

Deep-well injection may be the most cost effective alternative for inland treatment facilities, however, it can be expensive due to the costs of drilling, monitoring, and regulatory compliance, the latter of which may require a redundant well at a particular site (Acquaviva *et al.*, 1997). In addition, deep-well disposal of industrial wastes (including membrane concentrate and backwash) is not allowed in many states (Mickley, 2001).

When injecting into wells, the following parameters must be monitored (Brkic, 2003; Acquaviva *et al.*, 1997; US EPA, 2003):
1. Concentrate properties (density, viscosity, solids etc.)
2. Concentrate volume

3. Pump rates
4. Injection pressures
5. Dynamic wellhead pressures
6. Shut-in pressures
7. Pressure decline rates
8. Concentrate temperature
9. Corrosion effects
10. Wear of metal components

According to US EPA, Florida and Texas are the only states where the geologic conditions are considered suitable for deep well injection with strict regulatory controls and monitoring (US EPA, Region 4 and Region 5). Appendix A-3 presents case studies of plants that discharge concentrate to a deep well. These plants are located in Texas and Florida. A backup method of disposal must be available during periodic maintenance and testing of the integrity of the wells.

## *Land Application*

Land application of concentrate involves use of spray irrigation, infiltration trenches, and percolation ponds as well as irrigation of salt-tolerant crops and grasses (e.g., grasses used on golf courses, lawns, parks). Land application is typically used for smaller volumes of concentrates. Land application methods for concentrate disposal consist mainly of percolation ponds and spray irrigation systems. Percolation ponds or rapid infiltration basins are a viable disposal alternative where the waste will not significantly affect the quality of the groundwater in the receiving area. This option may be employed for discharge over shallow brackish aquifers, usually on islands or in areas which border estuaries or tidal creeks (Acquaviva *et al.*, 1997). Spray irrigation can be used for watering lawns, parks, or golf courses, and for preservation and enlargement of greenbelts and open spaces (Mickley, 2001), where use of concentrate for irrigation is limited by the quality and volume of the waste stream, climate, soil uptake rates, and the salt tolerance of the plants. Golf-course grasses and citrus trees require chloride concentrations less than approximately 1,000 mg/L (Pontius, 1997), and the presence of a high amount of sodium or trace elements can render reject brine unsuitable for irrigation purposes. However, brine with high divalent cation concentration may be useful in soil amelioration (Ahmed, 2000). Some halophytes that can tolerate water salinity up to 35,000 mg/L can be used for production of oil seeds or grains (Van der Bruggen and Vandecasteele, 2003). Use of concentrate for irrigation may require blending with treated municipal wastewaters to decrease the salinity to an acceptable range while increasing the nutrient load of the delivered water (Mickley, 2001). This method is usually used only for small concentrate flow rates, and an NPDES permit may be required for spray irrigation if the potential exists for runoff to reach another water body (Mickley, 2001). Messner *et al.* (1999) reported that concentrate from a nanofiltration plant was successfully blended with treated municipal wastewater and used for irrigation of two near-by golf courses. Monitoring of background groundwater at the golf courses indicated that it was not being influenced by reuse flow composition.

Environmental factors to be considered for land application of concentrate include (US EPA, 2003):

1. Availability of land
2. Availability of dilution water
3. Proximity of land to sensitive receptors
4. Soil conditions and percolation rates
5. Tolerance of vegetation to salinity
6. Sodium adsorption ratio (SAR) of soils
7. Trace metals uptake by soils
8. Depth groundwater table
9. Climate (land application may not possible year around)

Currently, Florida is the only state that currently uses land application for concentrate disposal (Hoepner, 2002). Usually dilution of the concentrate is required to meet groundwater standards. Where salinity levels are excessive, special salt tolerant species (halophytes) could be considered for irrigation. Land application also includes the use of percolation ponds and rapid infiltration basins. Use of concentrate for dust suppression, roadbed stabilization, and soil remediation has been used occasionally for small volume of concentrate (Mickley, 2001).

*Evaporation Ponds*

Instead of processing the concentrate by thermal evaporators (i.e., crystallizer or spray dryer), the concentrated brine may be discharged to evaporation ponds. Evaporation ponds can be feasible for small flows in relatively warm, dry climates with high evaporation rates, level terrain, and low land costs (Mickley, 2001). An impervious lining is required to prevent percolation of the concentrate to subsoils and into the groundwater. Evaporation ponds have an extensive history of use, are easy to construct, and require less maintenance and operator attention than mechanical systems (Ahmed *et al.*, 2000). At this time, evaporation ponds are probably the most widespread method of brine disposal for inland-based desalination facilities worldwide (Glater and Cohen, 2003).

Solar ponds are similar to evaporation ponds except that they are also used as heat sources for multistage flash evaporator units. Hence, power generation and thermal desalination are coupled with brine disposal (Glater and Cohen, 2003). Alternatively, power generated from the solar ponds can be used to pressurize the feed stream of membranes. Solar ponds combine solar energy collection with long-term storage and can provide reliable thermal energy at temperature ranges from 50 $^{\circ}$C to 90$^{\circ}$C (Lu *et al.*, 2001).

Design and operation considerations for the evaporations ponds include (INEEL, 2001):

1. Operating and maintenance procedures with monitoring and inspection plans,
2. Hydrologic report which includes sufficient information on the site's topography, soils, geology, surface hydrology, and groundwater hydrology,
3. Dike protection and structural integrity,
4. Leak detection,
5. Liner inspection procedures and compatibility evaluation,
6. Freeboard and overtopping prevention,
7. Nuisance and hazardous odor prevention,

8. Emergency response plan,
9. Type of waste stream (including chemical analysis),
10. Climatological factors including freeze/thaw cycles.

Monitoring wells are needed to ensure that the groundwater is not impacted by the salt levels in the evaporation pond. The salt precipitating in the pond should be removed periodically and can be either disposed at a landfill or used beneficially. Some beneficial uses of the recovered salts may include $CaCO_3$ in the cement industry, $MgSO_4$ in the ceramic industry, NaCl in the chlor-alkali industry, potassium for use in fertilizers, and lithium in the light metal industry (Ohya et al., 2001). Recovered salts may also be used in the production of steel, paper, fertilizers, and glass. Ahuja and Howe (2005) summarized the value of these recoverable resources and concluded that there might be an economic benefit if technologies can be implemented to obtain salts for specific industrial uses.

## Zero Liquid Discharge (ZLD) and Near ZLD

Zero liquid discharge (ZLD) systems typically use thermal evaporators, crystallizers and spray dryers to reduce concentrate to a solid product for landfill disposal. ZLD utilizes multiple effect evaporators and/or vapor compression evaporators to concentrate membrane residuals (Sethi et al., 2005; Bond et al., 2005). Electrodialysis can also be used to treat RO concentrate to increase TDS from 30,000 mg/L to 80,000 mg/L and evaporation can be used for further concentration up to 300,000 mg/L based on solubility limitations. These processes produce additional product water by recovering high-purity distillate from the concentrate stream, thereby reducing the size of the required membrane system and thus the amount of concentrate produced (Mickley, 2001). Energy is added to the system, often through the use of heat exchangers, to evaporate water from the brine, resulting in concentration of salt crystals. Energy requirements for the evaporation processes are significant. Therefore, zero liquid discharge can be considered for areas where surface water, sewer disposal, and deep well injection may not be feasible.

Appendix A-4 presents case studies for zero liquid discharge and near ZLD. Intermediate chemical precipitation for concentrate minimization demonstrated stable operation for brackish water with a moderate TDS level, and has been shown to be conceptually viable through the pilot testing for the City of North Miami Beach's Norwood Water Treatment Plant (Appendix A-4). ZLD and near ZLD processes may be feasible from sustainability perspective. Development of effective recovery and reuse options for the salt precipitated from the concentrate may allow these options to be feasible in the near future.

References

Acquaviva, P. G., Westrick, J. D., Dohme, C. I., and Derowitsch, R. W. (1997). Reverse Osmosis Concentrate Disposal Alternatives for Small and Medium Sized Systems in Southwest Florida. Membrane Technology Conference Proceedings, AWWA, pp. 961-977, February 23-26, New Orleans, LA.

Ahmed, M., W. H. Shayya, and D. Hoey (2000). Use of Evaporation Ponds for Brine Disposal in Desalination Plants. *Desalination* 130:155-168.

Ahuja, N., and K. J. Howe (2005). "Strategies for Concentrate Management from Inland Desalination." *Proceedings of the 2005 AWWA Membrane Technology Conference*, Phoenix, AZ, March 6-9, 2005.

Alameddine, I., El Fadel, M., and Mezher, T. (2007). Brine discharge from desalination plants: A modeling approach to an optimized outfall design, *Desalination*, 214: 241-260.

Bergman, R. A. (1995). Membrane Softening versus Lime Softening in Florida: A Cost Comparison Update, Desalination, 102 (1): 11-24.

Bond, R., Veerapaneni, S., and Edwards-Brandt, J. (2005). Reducing costs of inland desalination treatment. Journal of American Water Works Association, 97(3): 56-60.

Brkic, V. (2003). Waste disposal by deep well injection. Presented at the Exploration and Production Environmental Conference, March 10-12, San Antonio, TX.

El Fadel, M. and Alameddine, I. (2005). Desalination in arid regions: Merits and concerns, Journal of Water Supply: Research & Technology - AQUA, 54 (7): 449-461.

Gacia, E ., Invers, O., Manzanera, M., Ballesteros, E. and Romeo, J . (2007). "Impact of the brine from a desalination plant on a shallow seagrass (Posidonia oceanica) meadow", Estuarine Coastal and Shelf Science, 72, 579-590.

Glater, J. and Cohen, Y. ( 2003). Brine Disposal from Land Based Membrane Desalination Plants: A Critical Assessment, Prepared for the Metropolitan Water District of Southern California CEC PIER II – Contract No. 400-00-013.

INEEL (Idaho National Engineering and Environmental Laboratory) (2001). Evaporation pond sizing with water balance and make-up water calculations. Engineering design file.

Kimes, J. K. (1995). The Regulation of Concentrate Disposal in Florida. Desalination 102:87-92.

Lu, H., J. C. Walton, and A. H. P. Swift (2001). Desalination coupled with salinity-gradient solar ponds. Desalination 136:13-23.

Mahi, P. (2001). "Developing Environmentally Acceptable Desalination Projects." Desalination 138:167-172.

Malaxos, P. J. and Morin, O. J. (1990). Surface water discharge of reverse osmosis concentrates, Desalination, Volume 78, Issue 1, July 1990, Pages 27-40.

Malmrose, P., J. Lozier, M. Mickley, R. Reiss, J. Russell, J. Schaefer, S. Sethi, J. Manuszak, R. Bergman, and K. Atasi (2004). Committee Report: Current Perspectives on Residuals Management for Desalting Membranes. Journal AWWA 96:73-87.

Mercer, K.L. (2009). "Chemical treatment of high pressure membrane concentrate for improved residuals management", PhD Thesis, University of Massachusetts, Amherst, MA. http://scholarworks.umass.edu/dissertations/AAI3336976

Messner, S., Hart, G., Netzel, J. and Dietrich, J.A. (1999). Membrane Concentrate Reuse By Controlled Blending, Florida Water Resources Journal, 23-29, January.

Mickley, M. (2001). "Membrane Concentrate Disposal: Practices and Regulations," U.S. Bureau of Reclamation Desalination and Purification Research Development Program, Report No. 69, Denver, Colorado, USA.

Mickley, M. C. (2004). Membrane Concentrate Disposal: Practices and Regulation, Desalination and Water Purification Research and Development Program Report No. 19, U.S. Department of Interior, Bureau of Reclamation.

Ohya, H., Suzuki, T. and Nakao, S. (2001). Integrated system for complete usage of components in seawater : A proposal of inorganic chemical combination on seawater, Desalination, Volume 134, Issues 1-3, 20, 29-36

Sethi, S., Zacheis, A., and Juby, G. (2005). State-of-science and emerging and promising technologies for brine disposal and minimization for reverse osmosis desalination. In ACE 2005, AWWA Annual Conference and Exposition, San Francisco, California, 12-16 June 2005.

Skehan, S. and Kwiatkowski, P.J. (2000). Concentrate disposal via injection wells permitting and designconsiderations, Florida Water Resources J., May, 19-21.

Squire, D., Murrer, J., Holden, P., and Fitzpatrick, C. (1997). Disposal of reverse osmosis membrane concentrate. Desalination, 108(1-3): 143-147.

Tansel, B. and Sosnikhina, I. (2009). "Cost Comparison of Membrane Treatment and Concentrate Management Practices at Drinking Water Treatment Plants in Florida," Proceedings of the World Environmental & Water Resources Congress 2009, Kansas City, Missouri, May 17-21, 2009.

Truesdall, J., M. Mickley, and R. Hamilton (1995). Survey of Membrane Drinking Water Plant Disposal Methods. Desalination 102:93-105.

Tsiourtis, N. X. (2001). Desalination and the Environment. Desalination 141:223-236.

Tularam, G. A. and Ilahee, M. (2007). Environmental Concerns of Desalination Plants. J. *Environmental. Monitoring*, 9, 805-813.

US EPA (2003). United State Environmental Protection Agency, Membrane Filtration Guidance Manual. EPA 815-D-03-008.

Van der Bruggen, B. and Vandecasteele, C. (2003). Removal of pollutants from surface water and groundwater by nanofiltration: overview of possible applications in the drinking water industry, *Environmental Pollution*, Volume 122, Issue 3, 435-445.

Younos, T. (2005). Universities Council on Water Resources Journal of Contemporary Water Research and Education Issues, 132, 11-18, December.

# Chapter 5 - Economic Evaluation

Conrad G. Keyes, Jr., Emeritus Professor and Department Head, New Mexico State University, 801 Raleigh Road, Las Cruces, NM 88005, cgkeyesjr@q.com

Michael P. Fahy, El Paso Water Utilities, El Paso, TX 79961

Introduction

Some Economic Evaluations (only provided if the Owners of the Appendix A facilities have authorized such) have been provided. Historically (NRC, 2008), the relatively high financial costs of desalination has constrained the use of desalination technologies in many specific circumstances, but the cost picture has changed in a number of important ways. *There have been significant reductions in membrane costs and in other components of cost in the production of desalinated water. Perhaps, more significantly, the costs of other alternatives for augmenting water supplies have continued to rise along with the degree of treatment required of existing supplies, making desalination costs more attractive in a relative sense. A continuation of these trends would likely make desalination costs more attractive and less of a constraint in the future. The trend of desalination process cost reduction may be abetted through a program of strategically directed research aimed at achieving potentially large cost reductions.*

National Research Council Economic Recommendations

The following recommendations were suggested in chapter 6 of the NRC (2008) report.

> ***Substantial reductions in the financial cost of desalination will require substantial reductions in either energy costs or capital costs.*** *Energy and capital costs are the two largest components of financial cost for both thermal and membrane seawater desalination processes. It is important to recognize that reductions in scale or in the capital costs of a facility will have associated reductions in interest costs. In most instances, interest costs will be a large component of total costs. Future trends in energy costs will also be important inasmuch as significant increases in energy prices could offset or more than offset cost reductions in other areas and make desalination technologies less attractive.*
>
> ***For brackish water desalination, the costs of concentrate management can vary enormously from project to project and may rival energy and interest costs as the largest single component of cost.*** *The high cost of concentrate management at some inland locations ultimately offsets the cost advantage that can be obtained from utilizing feedwaters with lower salinity.*

*There are small but significant efficiencies that can be made in current membrane technologies that will reduce the energy needed to desalinate water and therefore offer potentially important process cost reductions.* Today's best available seawater RO membranes are operating at pressures that are only 40 percent greater than the osmotic pressure of seawater and therefore are approaching the theoretical limits of energy efficiency for membrane desalination. However, development of membranes that operate effectively at lower pressures could lead to 5 to 10 percent reductions in annual costs of desalinating seawater associated with a 15 percent decrease in energy use.

*Extending membrane life is likely to have a very small impact on desalination costs.* Today's best-available seawater RO membranes routinely operate for 5 or more years before needing to be replaced. The ability to extend membrane life past 5 years to 10 years will have a minimal impact on total costs given the small contribution of membrane replacement costs to total costs over a 5-year lifetime. However, the prevention of catastrophic failure is especially important because membrane failure within the first year of operation can cause an annual cost increase of over 25 percent. Future research efforts should be focused on mistake-proof, robust prefiltration to ensure against premature failure of the RO membranes.

*The costs of producing desalinated water have fallen in recent years but may rise in the future if the price or cost of energy rises faster than cost decreases from technological improvements.* Increases in energy costs lead disproportionately to increases in desalination costs and in the costs of transporting water long distances. The ultimate size of these increases, however, may be limited when the costs of fossil fuels reach the costs of other energy technologies, especially renewable energy technologies that can substitute for fossil fuels. Consequently, energy costs will not rise indefinitely even if possible fuel prices do rise more or less indefinitely. In considering the implications of increasing energy costs, it is important to recognize that alternative supply measures that also have high energy demands will be sensitive to future energy prices.

*Conservation and transfers from low- to high-valued uses will usually be less costly than supply augmentation schemes, including desalination.* In many circumstances, low-cost methods of demand management could provide significant water savings. Low-cost demand management techniques have not been exhausted and, so long as potential remains, demand management will offer the possibility of freeing up water to serve new uses at lower cost than desalination. Similarly, market-like transfers of water can also offer relatively low-cost ways of acquiring additional supplies of water. This is particularly true where additional water supplies are needed to support urban growth and where agricultural water is available for reallocation.

*Conservation and efficiency improvements that reduce the total demand for water often come with associated benefits (such as reduced energy costs), require little capital investment, and can be implemented relatively quickly. Ultimate costs will vary depending on the local details of water use, water available for transfer, previous efforts to improve efficiency, financial perspectives, and institutional factors that encourage or discourage different water policy choices.*

***To make the true costs transparent, the economic costs of desalination should be accounted for and reported accurately.*** *Failure to price water accurately can lead to inefficient use and overuse. Melded pricing or average cost pricing is frequently used pursuant to law or to address equity consideration. This practice understates the cost of desalinated water to the consumer, and the supplier should take care in publicly reporting the true and accurate economic costs.* (Reproduced with permission from the National Academies Press, Copyright 2008, National Academy of Sciences)

Summary of Southern California Technologies

A table (mainly Table 6) of the Southern California Regional Brine – Concentrate Management Study, Phase I Executive Summary (Bureau of Reclamation, 2009), provides a summary of the brine-concentrate treatment technologies and disposal options in that region of the United States.

*An assessment of the applicability to wastewater and groundwater sources for each technology was provided. The relative performance of the treatment and disposal options is rated based on the performance, amount of water recovered, water quality produced, design flexibility and implementability, technology footprint, amount of waste minimization, hazardous wastes/environmental concerns, chemical usage/handling and safety, proven technology, regulatory complexity, maintenance and labor requirements, aesthetics and public acceptance, and ease of use. These criteria were used to summarize the advantages and disadvantages of each technology.*

Typical Economic Summary (provided by Michael Fahy)

Some Economic Evaluations (only provided if the Owners of the Appendix A facilities have authorized such) have been provided; but the information isn't necessarily comparable as indicated by the two summaries as provided above. A typical economic summary (Table 5-1) was provided by Michael Fahy of El Paso Water Utilities for the Kay Bailey Hutchison Desalination Plant in El Paso, Texas. This plant uses the Deep Well Injection system and has been reported under Appendix A of this document.

Table 5-1. Typical Economic Summary Table by Project

| Project Name: Kay Bailey Hutchison Desalination Plant<br>Owner: El Paso Water Utilities (EPWU)<br>Location: El Paso, Texas<br>Membrane Process: Reverse Osmosis<br>Concentrate Disposal Method: Deep Well Injection | | | |
|---|---|---|---|
| | Capital Cost ($) | O&M Costs ($/year) | Amortized Unit Costs ($/Acre-Foot) |
| Treatment | $ 72 Million | $4,626,000 | 485 |
| Disposal | $ 19 Million | $200,000 | 49 |
| Total | $ 91 Million | $4,826,000 (assumes-80% operation) | 534 (per acre-foot of potable water) |
| Source of Construction Funds: Total of $91 Million from a variety of sources:<br>a. EPWU Bonds and Cash<br>b. Congressional Appropriation<br>c. Loan from Texas Water Dev. Board<br>d. U. S. Army Contribution | | | |

Source: Table by Michael Fahy with permission from El Paso Water Utilities, El Paso, TX

References

Bureau of Reclamation (2009). U.S. Department of the Interior, Southern California Regional Brine-Concentrate Management Study – Phase I, Lower Colorado Region, October

National Research Council (2008). "The Costs and Benefits of Desalination", *Desalination: A National Perspective,* National Academy of Sciences, p. 147-149, http://www.nap.edu/catalog/12184.html

# Chapter 6 – Implementation/Case Studies

Conrad G. Keyes, Jr., Emeritus Professor and Department Head, New Mexico State University, 801 Raleigh Road, Las Cruces, NM 88005, cgkeyesjr@q.com

Michael P. Fahy, El Paso Water Utilities, El Paso, TX 79961

Berrin Tansel, Professor, Florida International University, Civil and Environmental Engineering Department, Miami, FL 33174, tanselb@fiu.edu

Introduction

This chapter refers to case studies for the various types of concentrate disposal methods considered or implemented (or final disposal options) for sample desalination projects. It cites the brief information from the authors for the case studies by four different categories of concentrate disposal for respective concentrate projects, or plants.

Final Disposal Options Used

Final disposal options are concentrate management options that require no additional treatment or management. Final disposal options include concentrate discharge into oceans or bays; deep well injection, land application, or evaporation pond disposal; discharge to sanitary sewers or surface waters; or zero liquid disposal (ZLD), or near ZLD.

Each of these concentrate management technologies requires some regulatory approval prior to discharge in all parts of the United States.

Summary of Case Studies in Appendix A

Table 6.1 presents a summary of the brine-concentrate treatment technologies and discharge option provided in Appendix A of this report. As indicated in Chapter 1, the subcommittee chairs and their members reviewed the material provided by the authors listed below in Table 6.1. The names of the respective projects (or plants) are also provided in this summary.

In order to arrive at similar information about each project or test facility, the details were developed by each case study author as shown in Appendix A. The suggested format as originally developed by Ken Mercer is listed below in Table 6.2.

**Table 6.1 – Summary of Concentrate Management in Desalination Case Studies Provided**

A-1 Ocean and Bays Disposal (Subcommittee Chair – Jim Jensen)
    Nikolay Voutchkov – Carlsbad, CA plant
    Val S. Frenkel – Marin Municipal WD plant, San Rafael, CA
    Val S. Frenkel – Charles Meyers WW plant, Santa Barbara, CA

A-2 Sanitary Sewer or Surface Water Disposal (Subcommittee Chair – Harold Thomas)
    Michael Fahy, John Balliew, & Anthony Tarquin – Pilot Research of Non-Irrigation Season Flows to River
    Khalil Atasi and Colin Hobbs – Ormond Beach, FL (dual – land application/sewer)
    Berrin Tansel - Joe Mullins RO plant

A-3 Deep well Injection, Land Disposal, and Evaporation Ponds (Subcommittee Chair – Ken Mercer)
    Michael Fahy, Scott Reinert & Kenneth Mercer - EPWU Kay Bailey Hutchison Plant
    Berrin Tansel - North Collier Regional plant
    Berrin Tansel – Melbourne, FL plant
    James Jensen – Dalby Stage 2 plant, Queensland, Australia

A-4 Zero Liquid Discharge (ZLD) and Near ZLD (Subcommittee Chair – Sandeep Sethi)
    Berrin Tansel - NASA Closed Loop
    Sandeep Sethi – South Florida WMD

**Table 6.2 - Layout of Case Study Template**

    <u>Brief Description of Project</u>
    Management Approach:
    Committee Member(s):
    Project Contact(s):
    Project Name:
    Project Location:
    Desalination Process:
    WTP Information:
    Rated Capacity:
    Max. Concentrate Flow:
    Typical Production:
    Typ. Concentrate Flow:

Abstract
Process Design and Configuration
Figure 1: Schematic of Pilot Treatment System
Project background:
Description of the proposed solution:
Data Collection Procedures:
Permitting and regulatory overview and procedure:
Analyses of plant concentrate:
Economic Evaluation:
Key project lessons learned:
Pictures
Acknowledgements
Reference(s)

The summary of case studies represented in Table 6.1 shows the diverse categories of disposal options. The individual case studies in Appendix A provide some detail regarding the advantages and disadvantages of each type of disposal method, including some economics for the technology selected.

# Appendix A-1

# Oceans and Bays Discharge Case Studies

Proposed Carlsbad Seawater Desalination Plant – Carlsbad, CA
Nikolay Voutchkov

Marin Municipal Water District (MMWD) Desalination Project - San Rafael, CA
Val S. Frenkel

The Charles Meyer Desalination Facility - Santa Barbara, CA
Val S. Frenkel

## ASCE/EWRI Task Committee
## CONCENTRATE MANAGEMENT IN DESALINATION
### Case Study

Management Approach: - Ocean Discharge - Collocation of Desalination Plant and Power Plant Discharges

Committee Member(s): Nikolay Voutchkov

Project Contact(s): Peter MacLaggan, pmaclaggan@posedion1.com

Project Name: Proposed Carlsbad Seawater Desalination Plant

Project Location: Carlsbad, California

**Desalination Process**: Source seawater collection from the cooling water discharge of the Encina Power Generation Station. Pretreatment is by conventional single stage, dual media gravity filtration (sand and anthracite); and source seawater conditioning by pH adjustment and antiscalant addition. Reverse osmosis separation is by a single-pass, single stage membrane desalination system. Product water permeate post-treatment by calcite contact filtration, pH adjustment, addition of carbon dioxide and chlorination by sodium hypochlorite and ammonia. No product water fluoridation is planned at this time. However the desalination plant is designed to accommodate the installation of fluoridation facilities, if such facilities are needed in the future.

WTP Information:
- Rated Capacity: 50 MGD
- Max. Concentrate Flow: 54 MGD
- Typical Production: 48 to 50 MGD
- Typ. Concentrate Flow: 48 to 50 MGD

### *Abstract*
The Carlsbad seawater desalination plant will be collocated with the Encina Power Generation Station, located in the City of Carlsbad. The desalination plant will use the existing power plant outfall to mix the concentrate discharge with the cooling water discharge from the power plant and will use the buoyancy of the warm power plant discharge to balance the negative buoyancy of the desalination plant concentrate discharge and accelerate its mixing.

### *Process Design and Configuration*
The desalination plant will not have a separate intake and discharge. Instead the desalination plant will use the existing cooling water discharge facilities of the Encina Power Generation Station to collect seawater for fresh water production and to convey the desalination plant discharge to the ocean (see Figure 1). The

desalination plant treatment will incorporate single-stage granular media filtration followed by 20-micron cartridge filtration, single-stage seawater reverse osmosis (SWRO) membrane separation and post treatment through calcite contact tanks and chlorination. The plant will operate at 45 to 55 % recovery and energy recovery from plant concentrate will be completed using pressure-exchangers.

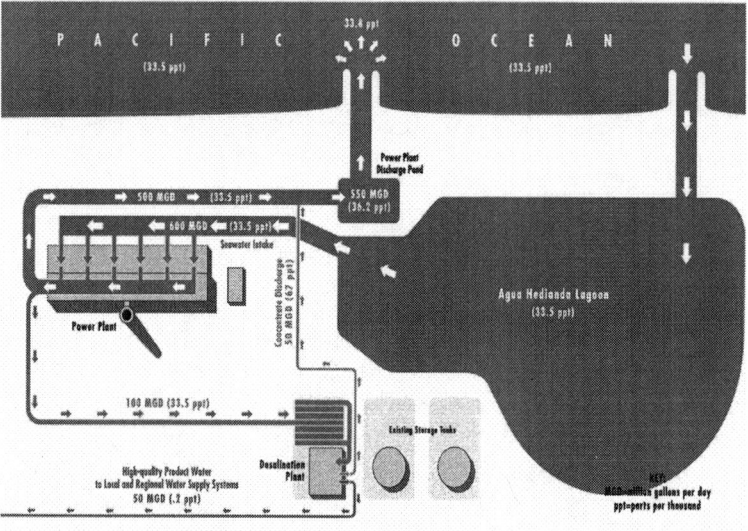

Figure 1. Schematic of Carlsbad Desalination System (Diagram by Nikolay Voutchkov 2004, with permission from Water & Wastewater.com)

*Project background:*
Currently, San Diego County relies on water imported from Colorado River and the Bay Delta for approximately 90 % of their drinking water supply. Once in operation, the 50 MGD Carlsbad seawater desalination plant will provide approximately 8 % of the total County water supply. This plant will serve over 350,000 residents of the City of Carlsbad; Valley Center, Rincon del Diablo, Olivenhain, Vallecitos and Rainbow Municipal Water Districts; as well as Sweetwater Authority and Santa Fe Irrigation District. The City of Carlsbad plans to supply 100 % of their drinking water from the desalination plant and to discontinue the use of imported water. The other utilities will supplement between 5 and 15 % of their current water supplies with desalinated seawater. The project will provide local draught-proof water supply and will reduce the need for water exports to San Diego County from Colorado River and States Water Project.

As shown on Figure 1, under typical operational conditions approximately 600 MGD of seawater enters the power plant intake facilities and after screening is pumped through the plant's condensers to cool them and thereby to remove the waste heat created during the electricity generation process. The Carlsbad desalination plant intake structure is connected to the end of this discharge canal and under normal operational conditions would divert 100 MGD of the 600 MGD of cooling water for production of fresh water. Approximately 50 MGD of the diverted cooling seawater would be converted to fresh drinking water via reverse osmosis membrane separation. The remaining 50 MGD would have salinity approximately two times higher than that of the ocean water (67 ppt vs. 33.5 ppt). This seawater concentrate would be returned to the power plant discharge canal downstream of the point of intake for blending with the remaining cooling water prior to conveyance to the Pacific Ocean. Under average conditions, the blend of 500 MGD of cooling water and 50 MGD of concentrate would have discharge salinity of 36.2 ppt, which is within the 10 % natural variation of the ocean water salinity (36.9 ppt) in the vicinity of the existing power plant discharge.

***Description of the proposed solution:***
Concentrate generated at the seawater desalination plant will be blended with warm cooling water from the Encina power generation station and the blend will be discharged to the Pacific Ocean through the existing power plant ocean discharge structure, which extends approximately 700 feet from the shore (see Figure 2). The concentrate TDS concentration is projected to be maintained at 40 ppt or less at all times.

Figure 2. Carlsbad Seawater Desalination Project (Photograph by Nikolay Voutchkov 2004, with permission from Water & Wastewater.com)

***Data Collection Procedures:***
Series of concentrate and source water quality data were collected using the desalination demonstration plant installed at the Encina Power Generation Station. In addition, the project proponent, Poseidon Resources, has completed salinity tolerance study, which was used by the San Diego Regional Water Quality Control Board to establish the monthly average and maximum concentrate salinity concentration limits for this project – 40 ppt and 44 ppt, respectively. The plant concentrate management system will be designed to maintain maximum discharge salinity concentration of 40 ppt at all times in order to provide a 10% safety margin as compared to the regulatory maximum plant discharge limit of 44 ppt established by the plant permit.

***Permitting and regulatory overview and procedure:***
The environmental review process for this project included four key components; Environmental Impact Report (EIR), certified by the City of Carlsbad; NPDES discharge permit approved by the San Diego Regional Water Quality Control Board; Coastal development permit issued by the California Coastal Commission, and the site lease for plant intake and outfall provided by the California State Lands Commission.

Construction and operation of a desalination plant requires obtaining permits or approvals from a number of regulatory agencies including:

- Compliance with CEQA – City of Carlsbad was the CEQA Lead Agency
- National Pollutant Discharge Elimination System (NPDES) Permit from San Diego RWQCB
- Intake and Discharge Lease from California State Lands Commission.
- Coastal Development Permit from the California Coastal Commission.
- Consultation with the U.S. Fish and Wildlife Service (USFWS) and NOAA Fisheries in accordance with Section 7 of the Federal Endangered Species Act (ESA)
- Consultation with CDFG through California Fish and Game Code Section 2081 for state listed threatened or endangered species
- Incidental Harassment Authorization (IHA) under the Marine Mammal Protection Act for potential impingement of marine mammals on the screens of the power plant in case the power plant discontinues the use of once-through cooling for its generation facilities and the desalination plant becomes the sole user of the existing power plant intake.

***Analyses of plant concentrate:***
Plant concentrate is projected to have salinity of 60 to 67 ppt. The dilution of the combined concentrate/effluent discharge was modeled using a tri-dimensional hydrodynamic model developed by the Scripps Institute of Oceanography. This CDF model used for these projects was initiated for four sets of average and extreme environmental conditions: 1) a worst-case day; 2) an average day; 3) a worst-case month, and 4) an average month. For each of these conditions the model analysis

studied dispersion and dilution of the concentrate for 50 MGD desalinated water production. The key variables controlling the concentration and persistence of the elevated salinities and temperatures associated with the combined power plant-desalination plant discharge used in these models are: (1) desalination plant and power plant flow rates; (2) ocean salinity; (3) ocean temperature; (4) ocean water levels; (5) mixing wave action; (6) mixing current action; and (7) mixing wind action.

Aquatic Life Salinity Tolerance Threshold was determined as a part of the concentrate environmental studies. All marine organisms are naturally adapted to changes in seawater salinity. These changes occur seasonally and are mostly driven by the evaporation rate through the ocean surface, by rain/snow events and by surface water discharges. The natural range of seawater salinity fluctuations could be determined based on information from sampling stations located in the vicinity of the discharge and operated by federal, state or local agencies and research centers responsible for ocean water quality monitoring. Typically, the range of natural salinity fluctuation is at least +/- 10 % of the average annual ambient seawater salinity concentration. For example, based on ocean water quality monitoring data collected for the development of environmental impact reports for the Carlsbad Desalination Project the average annual ocean water salinity (i.e., TDS) concentration is 33,500 mg/L, while the maximum salinity reaches 36,800 mg/L, which is approximately 10 % above the average. The "10 % increment above ambient ocean salinity" threshold is a conservative measure of aquatic life tolerance to elevated salinity. The actual salinity tolerance of most marine organisms is significantly higher than this level. Detailed analysis and salinity tolerance studies performed for the Carlsbad desalination project indicate that the most species inhabiting the Southern California Bight, and especially these found in the discharge areas of the two plants, have a salinity tolerance threshold of 40,000 mg/L or higher. The Southern California Bight is an open embayment extending from Point Conception, California into Baja California, Mexico and 125 miles offshore.

*__Economic Evaluation:__*
A new 1-mile long outfall pipeline was considered as an alternative to the use of the existing open power plant outfall structure which extends approximately 700 feet from the shoreline. Construction of new outfall structure was not found viable because of the potential negative environmental impact on existing kelp beds. Concentrate discharge using subsurface wells was not found feasible due to the unfavorable hydro-geological conditions in the vicinity of the plant and the significant environmental impacts of construction and maintenance of numerous discharge wells along the shore.

*__Key project lessons learned:__*
Collocation provides number of cost and environmental benefits, which are not available through alternative intake and discharge structures such as new open ocean intakes and outfalls, and subsurface intakes. Avoidance of construction of intake and discharge facilities reduces the overall cost of water production by 10 to 20 %. Since

the power plant has already used the source seawater for cooling, the operation of the desalination plant does not create incremental impingement and entrainment of marine organisms. Under the collocation configuration, the power plant discharge serves both as an intake and discharge to the desalination plant. Four key benefits stem from this arrangement: 1) construction of a separate desalination plant outfall structure is avoided, thereby decreasing 10 to 20 % of the overall costs for seawater desalination; 2) salinity of the desalination plant discharge is reduced as a result of mixing and dilution of the membrane concentrate with power plant discharge, which has ambient seawater salinity; 3) because a portion of the discharge water is converted to potable water, total quantity of the power plant thermal discharge is reduced, which lessens negative effects of the power plant thermal discharge on the aquatic environment; 4) blending of the desalination plant and power plant discharges results in accelerated dissipation of both salinity and thermal discharges. One key additional advantage of collocation is the overall reduction of the desalination plant power demand and associated costs of water production as a result of warmer source water. The source water of the SWRO plant is typically 5-15°C higher than the temperature of the ambient ocean water. This is of significant benefit, especially for desalination plants with cold source seawater, because the RO membrane separation of 10°C of warmer seawater requires about 5-8% lower feed pressure, and therefore, a proportionally lower energy use for seawater desalination. Since the power costs are about 20-40% of total costs for production of desalinated water, use of warmer source water has a measurable beneficial effect on the overall water production costs.

## ASCE/EWRI Task Committee
## CONCENTRATE MANAGEMENT IN DESALINATION
### Case Study

Management Approach: Concentrate discharge to the wastewater plant outfall

Committee Member(s): Val S. Frenkel, Kennedy/Jenks Consultants

Project Contact(s): Bob Castle, Water Quality Manager, MMWD, (415) 945-1556

Project Name: Marin Municipal Water District (MMWD) Desalination Project

Project Location: MMWD-owned land near Pelican Way in the City of San Rafael, California.

WTP Information (PROPOSED):
- Rated Capacity: 5 MGD
- Max. Concentrate Flow: 5 MGD
- Typical Production: N/A
- Typ. Concentrate Flow: N/A

*Final Capacity: 15 MGD*  *Max. Concentrate Flow: 15 MGD*
*Initial Phase Production: 5 MGD*  *Concentrate Flow: 5 MGD*

### Abstract

Due to a number of factors, the current MMWD water demand exceeds the current reliable water supply by approximately 3,300 AFY. If water demand in MMWD increases as projected and if no new water supply is provided, this water supply deficit will increase to 6,700 AFY by the year 2025.

Although SWRO technology innovations have made implementation of desalination more cost feasible, current environmental concerns and regulatory requirements regarding SWRO brine discharges can impact permitting and operation of a full-scale desalination facility. Brine produced by a full-scale MMWD desalination facility will be mixed with the relatively low-salt wastewater effluent discharged by a co-located wastewater treatment plant. The mixture of brine from the desalination plant and wastewater effluent, referred to as whole effluent (WE), will be discharged to the Bay through an existing deep-water outfall. To ensure that the brine disposal process would meet the requirements of the San Francisco Bay Area Regional Water Quality Control Board and not adversely impact the aquatic environment of San Francisco Bay, MMWD conducted an ecotoxicity study with pilot plant WE. No significant effects on survival were observed among several aquatic indicator organisms exposed to WE. Minor impacts on sublethal endpoints among these species were observed at concentrations low enough that minimal receiving water dilution would eliminate them. In addition, the study results provide some evidence that suggest RO brine provides a protective function by mitigating toxicity occurring in wastewater effluents during periods where contaminants would be more concentrated.

## *Process Design and Configuration*

Intake would be directly from San Rafael Bay through screened pumps located at the end of a 2,000 foot long replacement to an existing wood pier.

Suspended solids would be removed by pretreating the raw water using a microfiltration/ultrafiltration (MF/UF) system. A strainer system with 100 micron nominal removal would be required ahead of the MF/UF filters to protect the membrane fibers from damage. The strainer system would be composed of a plastic compressed disc-type strainer. Periodic addition of a coagulant may be required ahead of the MF/UF filters to reduce the high levels of organics in the Bay source water and to reduce bio-fouling of the MF/UF and RO membrane elements.

Reverse osmosis would be followed by "post-treatment" of desalted water to produce drinking water with taste and other characteristics comparable to that currently provided for MMWD's customers. The concentrate from the RO process would be mixed with treated effluent from the Central Marin Sanitation Agency (CMSA) and discharged back into San Rafael Bay.

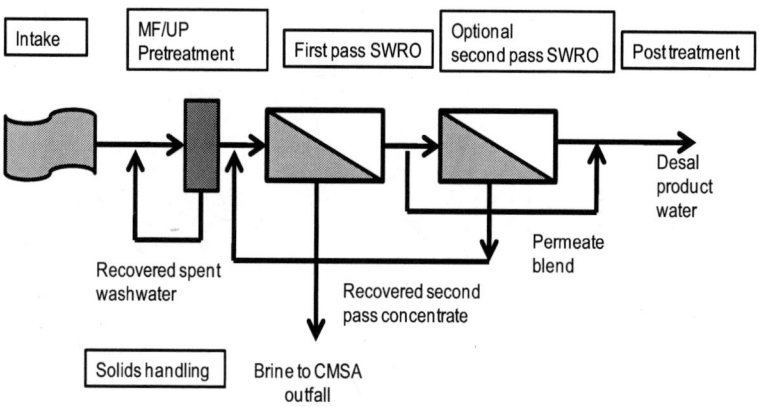

**Figure 1. Schematic of Pilot Treatment System (Courtesy of Val Frenkel).**

## *Project background:*

Due to a number of factors, the current MMWD water demand exceeds the current reliable water supply by approximately 3,300 AFY. If water demand in MMWD increases as projected and if no new water supply is provided, this water supply deficit will increase to 6,700 AFY by the year 2025.

MMWD is proposing a 5 million gallon per day (MGD) desalination plant to be located in the City of San Rafael, California. Certain infrastructure would be oversized so that it could accommodate potential expansion of the plant. In the event MMWD decides to expand the plant, expansion would occur in 5 MGD increments, up to a maximum capacity of 15 MGD – i.e., to 10 MGD, and to 15 MGD. "Raw water" or "feed water" from San Francisco Bay would be collected through an intake at the end of the proposed refurbished pier near the Richmond–San Rafael Bridge and subjected to various forms of treatment to produce drinkable (potable) water. The desalination process would convert about half the volume of raw water taken from the Bay into drinking water. The remaining water, or brine, would be discharged back to the Bay via an outfall operated by a municipal wastewater treatment plant. Blending of brine with treated wastewater effluent would reduce the concentration of dissolved salts in the brine prior to its release into San Francisco Bay.

Although new RO technology innovations have made implementation of desalination for drinking water more cost feasible, concerns over potential environmental impacts of an RO based desalination plant have arisen. Suspected adverse impacts to the aquatic environment associated with desalination include high salt concentrations, defouling and pretreatment chemicals, and heavy metals toxicity. Although diluting the pilot plant brine waste with treated wastewater prior to discharge can provide assurance that many of these impacts will be avoided, preliminary public comments solicited early in the CEQA permitting process indicated that environmental impacts of the MMWD desalination plant were still a concern.

### *Description of the proposed solution:*
As a component of the strategy to address potential environmental impacts and public perception, MMWD commissioned an ecotoxicity study as part of the CEQA environmental planning and permitting process. The study was designed to assess the potential for desalination discharge water to adversely impact the aquatic ecology of San Francisco Bay by employing standard EPA bioassay methods for assessing toxic effects using several native species and various trophic levels. WE samples used for the study were created by mixing RO waste brine from the MMWD pilot plant with treated wastewater effluent from the co-located wastewater treatment plant. Two WE blends were analyzed: a High Brine Blend representing dry weather discharges, and an Average Brine Blend representing normal flow discharges.

### *Data Collection Procedures:*
This two phase ecotoxicity study employed several aquatic indicator organisms. Two WE blends: High (47 ppt) Brine and Average Brine (16 ppt) were collected at the pilot plant. Organisms were exposed to the Average Brine blend and *two* High Brine blend treatments: 1.) Adjusted to protocol salinity levels (30 ppt), and 2.) Maintained at expected discharge level (47 ppt). The same dilution series for all three blend treatments (i.e. 5, 10, 25, 50 and 100%) was used. Bioassays were performed in accordance with EPA and/or ASTM approved methods (USEPA 1995, 2002a, and 2002b).

**Determination of Dilution Discharge Dilution.** The dilution of the combined concentrate/effluent discharge was modeled using U.S. Environmental Protection

Agency's Visual Plumes (VP) mixing zone model. Not enough data were available to perform hourly dilution modeling for a whole year, and so a statistical approach using the Monte Carlo method was chosen as the preferred methodology. Five-thousand scenarios were generated with the Monte Carlo method and run with VP. Key results included the average dilution that is achieved in the near-field, defined where the plume hits the bottom of the Bay floor or reaches the surface of the water. The plume would continue to dilute from ambient turbulence beyond this point, i.e., in the far-field.

***Phase I.*** Five bioassay test methods were employed for Phase 1 of the chronic screening study. The 48-hour giant kelp, *Macrocystis pyrifera,* germination and growth test and the 48-hour bay mussel, *Mytilus edulis*, percent survival was performed in accordance with *Short-term Methods for Estimating the Chronic Toxicity of Effluents and Receiving Water to Marine and Estuarine Organisms, West Coast Edition* (USEPA 1995). The 96-hr *Thallasiosira pseudonana* cell density test was performed in accordance with the *Annual Book of ASTM Standards* (ASTM 2003). The 7-day inland silverside, *Menidia beryllina,* and *Mysidopsis bahia* survival and growth tests were performed in accordance with *Methods for Measuring the Acute Toxicity of Effluents and Receiving Waters to Marine and Freshwater Organisms, Fifth Edition* (USEPA 2002a). The five bioassays were performed concurrently in one testing episode.

***Phase II.*** Three bioassay test methods were employed for the MMWDPP Phase 2 screening study. The 48-hour giant kelp, *Macrocystis pyrifera,* germination and growth test and the 48-hour bay mussel, *Mytilus edulis*, percent survival was performed in accordance with *Short-term Methods for Estimating the Chronic Toxicity of Effluents and Receiving Water to Marine and Estuarine Organisms, West Coast Edition* (USEPA 1995). The 7-day inland silverside, *Menidia beryllina*, survival and growth test was performed in accordance with *Methods for Measuring the Acute Toxicity of Effluents and Receiving Waters to Marine and Freshwater Organisms, Fifth Edition* (USEPA 2002a). The three bioassays were performed concurrently over two consecutive test episodes.

***Test Solution Preparation.*** Effluent samples for both Phase I and Phase II testing were collected and transported on ice to the WESTON laboratory under chain-of-custody. Samples were received at the laboratory, and after initial water quality measurements were taken and used immediately for testing. In order to achieve salinity levels consistent with biologically tolerant, EPA protocol recommended levels, the Average-brine Blend salinity for all tests was raised to 30 ± 2 ppt with CoralSea™ synthetic seasalts, and the High-brine Blend was diluted to 30 ± 2 ppt with spring water (Arrowhead™). An "Unadjusted" High-brine Blend was also tested without any salinity manipulation to asses the effects of higher salinity levels within the discharge mixing zone. The diluent and dilution control for all tests was filtered, UV treated seawater from Bodega Bay or San Francisco Bay. A salinity control made of deionized water and CoralSea™ synthetic seasalts at a salinity of 30 ± 2 ppt was included for all Average-brine Blends to eliminate the potentially confounding effects of the seasalts. The dilution series used for all adjusted Blends and the Unadjusted High-brine Blend treatment consisted of 5, 10, 25, 50, 75 and 100% treatments. An "Unadjusted" Average-brine Blend was also tested with the three species tolerant of lower salinity levels (*T. pseudonana, M. beryllina* and *M. bahia*) to assess directly the potential effects of average blend ratios within the discharge mixing zone. The Unadjusted Average-brine Blend tests were performed with undiluted sample only (i.e. no dilution series).

***Statistical Analysis.*** At the conclusion of each test, data were evaluated using the statistical program ToxCalc to determine the ECp and NOEC. ToxCalc is a comprehensive statistical application that follows standard guidelines for acute and chronic toxicity data analysis. Statistical effects can be measured by the ECp, the point estimate of the concentration at which an inhibitory effect is observed in p% of the organisms. The measured effect includes survival and reproductive/development endpoints in the tests. The No Observable Effect Concentration (NOEC) is the sample concentration at which the measured effects are not significantly different from those measured in the control. For the California Regional Water Quality Control Board, San Francisco Bay Region, in the case of an effect on an intermediate concentration, the next significantly different concentration below the lowest effect is considered the NOEC.

### *Study Results:*
Results are provided in Tables 1 through 3 below. Tables 1 and 2 show a summary of results for the three species that were determined to be more sensitive to WE in Phase I and exclusively tested in Phase II. As shown in the two tables, there were no effects on mortality among any of the species tested, and the observed sublethal effects would not generally be expected to occur with the minimum 10:1 discharge dilution provided in the discharge mixing zone. The two sublethal endpoints that did show NOEC values below 10% WE (Phase I kelp and Phase II, Episode 1 mussel), were coupled with much higher IC50 values, indicating the significant effects observed in the lower WE dilutions were only slight (i.e. relatively low absolute difference from the control). Table 3 shows expected and actual toxicity results observed with all test episodes for the final three species exposed to High Brine blend. The expected High Brine blend results are based on results of the Average Brine blend treatment and normalized to account for dilution necessary to lower the High Brine blend WE salinity to the protocol accepted range.

Table 1. Average Blend Results Summary (Weston Solution 2007, with permission from Weston Solution, Inc.)

| ENDPOINT | SPECIES | Phase 1 | Phase 2 Episode 1 | Phase 2 Episode 2 | MEANa (n=3) |
|---|---|---|---|---|---|
| Survival LC50 (%) | Giant kelp | >100 | >100 | >100 | 100 |
| | Inland silverside | >100 | >100 | >100 | 100 |
| | Bay mussel | >100 | >100 | >100 | 100 |
| Survival NOEC (%) | Giant kelp | 100 | 100 | 100 | 100 |
| | Inland silverside | 100 | 100 | 100 | 100 |
| | Bay mussel | 100 | 100 | 100 | 100 |
| Growth/Development IC50 (%) | Giant kelp | >100 | >100 | >100 | 100 |
| | Inland silverside | >100 | >100 | >100 | 100 |
| | Bay mussel | 61.3 | 33.9 | >100 | 65.1 |
| Growth/Development NOEC (%) | Giant kelp | 5 | 25 | 25 | 18.3 |
| | Inland silverside | 100 | 100 | 50 | 83.3 |
| | Bay mussel | 25 | <5 | 50 | 26.7 |

Table 2. High Blend Results Summary (Weston Solution 2007, with permission from Weston Solution, Inc.)

| ENDPOINT | SPECIES | Phase 1 | Phase 2 Episode 1 | Phase 2 Episode 2 | MEANa (n=3) |
|---|---|---|---|---|---|
| Survival LC50 (%) | Giant kelp | >100 | >100 | >100 | 100 |
| | Inland silverside | >100 | >100 | >100 | 100 |
| | Bay mussel | >100 | >100 | >100 | 100 |
| Survival NOEC (%) | Giant kelp | 100 | 100 | 100 | 100 |
| | Inland silverside | 100 | 100 | 100 | 100 |
| | Bay mussel | 100 | 100 | 100 | 100 |
| Growth/Development IC50 (%) | Giant kelp | >100 | >100 | >100 | 100 |
| | Inland silverside | >100 | >100 | >100 | 100 |
| | Bay mussel | >100 | >100 | >100 | 100 |
| Growth/Development NOEC (%) | Giant kelp | 50 | 100 | 75 | 75 |
| | Inland silverside | 100 | 100 | 100 | 100 |
| | Bay mussel | 100 | 25 | 100 | 75 |

**Table 3. Dilution vs. Fortification (Weston Solution 2007, with permission from Weston Solution, Inc.)**

| ENDPOINT | Average Blend (%) | Expected High Blend* (%) | Observed High Blend (%) | High Blend Observed Δ |
|---|---|---|---|---|
| Phase 1 Mussel IC50 | 61.3 | 96.2 | >100 | 3.8 |
| Phase 2.1 Mussel IC50 | 33.9 | 53.2 | >100 | 46.8 |
| Phase 1 Kelp NOEC | 5.0 | 10 | 50 | 40 |
| Phase 2.1 Kelp NOEC | 25 | 50 | 100 | 50 |
| Phase 2.2 Kelp NOEC | 25 | 50 | 75 | 25 |
| Phase 2.2 Silverside NOEC | 50 | 100 | 100 | 0 |
| Phase 1 Mussel NOEC | 25 | 50 | 100 | 50 |
| Phase 2.1 Mussel NOEC | 5 | 10 | 25 | 15 |
| Phase 2.2 Mussel NOEC | 50 | 100 | 100 | 0 |

ᵃCalculated by applying the dilution factor used to reduce the high blend salinity from 47 to 30 ppt using laboratory grade moderately hard water.

***Permitting and regulatory overview and procedure:*** Construction and operation of a desalination plant requires obtaining permits or approvals from a variety of resource agencies. Along with the CEQA compliance effort underway, the environmental permits or approvals that will likely be required for the proposed project include:

  National Pollutant Discharge Elimination System (NPDES) Permit from the San Francisco Bay RWQCB

  Section 401 Water Quality Certification from the San Francisco Bay RWQCB

  Section 404 Permit from the U.S. Army Corps of Engineers (USACE)

  Permit from the San Francisco Bay Conservation and Development Commission

  Permit to Operate from the Bay Area Air Quality Management District (BAAQMD)

  Consultation with the U.S. Fish and Wildlife Service (USFWS) and NOAA Fisheries in accordance with Section 7 of the Federal Endangered Species Act (ESA)

  Consultation with CDFG through California Fish and Game Code Section 2081 for state listed threatened or endangered species

Incidental Harassment Authorization (IHA) under the Marine Mammal Protection Act for potential disturbance to marine mammals during pile driving for construction of the pier.

To address the key CEQA objectives, the ecotoxicity study results were used to support the environmental impacts assessment included in the EIR and to establish the testing protocol to be adopted for the Regional Water Quality Control Board NPDES permit discharge permit for future pilot and full-scale plant discharges

### *Economic Evaluation:*
The proposed 5-MGD desalination facility would cost $115 million, which includes a new pier for intake and over $20 million for distribution system improvements, plus very conservative contingency factors. The cost of treated water would be between $2,000 and $3,000 per acre-foot.

### *Key project lessons learned:*
The MMWD ecotoxicity study demonstrated the following about desalination processes:

- Desalination can produce high quality and good tasting drinking water even from degraded source waters;
- Entrainment and impingement can be mitigated through good intake design;
- Desalination brine can be discharged without significant adverse effect to the marine environment;
- Exposure to neither Average nor High Brine blends in this case caused statistically significant mortality to any species tested;
- Sublethal effects were generally elicited by WE dilutions that were more concentrated than the dilution provided in the mixing zone.
- The High Brine effects observed on kelp spore germ-tube growth (Phase I and Phase II, Episode 2) and bivalve embryo development (Phase II, Episode 1) were substantially less significant than those observed with the Average Brine blend exposures.
- Unknown constituents in higher brine blends may have mitigating effect on WE toxicity, similar to results seen with Water Effects Ratio studies performed for wastewater dischargers.

## Pictures

**Photographs by Val Frenkel 2004, with permission from Marin Municipal Water District (MMWD)**

*Acknowledgement*

Scott Bodensteiner for providing the information from Weston Solution, Inc.

*References*

Marin Municipal Water District, 2007, DRAFT Environmental Impact Report (EIR) November, Marin Municipal Water District Desalination Project.

Weston Solution, Inc, 2007, "Marin Municipal Water District Desalination Pilot Plant Chronic Toxicity Screening Program", November, Tiburon, CA.

## ASCE/EWRI Task Committee
## CONCENTRATE MANAGEMENT IN DESALINATION
### Case Study

Management Approach: Discharge of Concentrate to Oceans

Committee Member(s): Val S. Frenkel, Kennedy/Jenks Consultants

Project Contact(s): Rebecca Bjork, Water Resources Manager,
City of Santa Barbara, (805) 897-1914

Project Name: The Charles Meyer Desalination Facility

Project Location: 525 E. Yanonali St., Santa Barbara, California, USA

Proposed Desalination Process:
    Original Capacity: 6.7 MGD (Ref. 4) Max. Concentrate Flow: 12.5 MGD
    Current Capacity: 2.8 MGD (Ref. 4) Max. Concentrate Flow: 3.4 MGD *
    Typical Production: 0 MGD** Typ. Concentrate Flow: 0 MGD**

\* Current Max. Concentrate Flow calculated here based on current capacity and on Ref 1: "Approximately 45% of the ... seawater ... becomes drinking water."
\*\* The plant was constructed in 1991-1992 (Ref. 1) but only operated for 3 months (Ref. 2), producing a total of 419 acre-feet (af) during start-up and testing (Ref. 4).

### *Abstract*
An existing wastewater outfall was selected as an obvious disposal method that would provide initial dilution and would minimize time and cost of construction. The plant is in long-term storage due to abundant rainfall since 1991 and reduced demand. Regional partners have dropped out and capacity has been sold off. The facility is now fully owned by the City of Santa Barbara (Ref. 4).

### *Process Design and Configuration (Figures 1 thru 3 attached)*
Ocean water is pumped at a very low pressure through a 2,500 foot seawater intake line [abandoned wastewater discharge line – relined with HDPE (Ref 4)] to the facility. The incoming seawater is pretreated in round horizontal media filters. Two sets of filters — primary, consisting of sand, gravel, and anthracite, and secondary consisting of the same media as primary, plus garnet. Cartridge filters complete the pre-treatment.

Pre-treated seawater is pressurized to 800 psi for a single pass through RO membranes. Approximately 45% of the pressurized seawater becomes drinking water (Ref. 1).

Modular desalination units were installed in trailers. Each "unit" consists of a pumping trailer and a membrane trailer. Each unit is capable of producing 625 afy of desalinated water (Figure 4 and Ref. 4).

Finished water is pumped into an existing water main for distribution to water customers. The concentrate is combined with treated wastewater from the adjacent wastewater treatment plant, and discharged to the ocean at the end of the 1.5 mile long outfall line (Ref 1). The outfall line includes a diffuser (Ref 3).

It takes approximately 6,600 kilowatt hours of electrical energy to produce one acre-foot (326,000 gallons) of desalted water (Ref. 1).

### *Project background:*

- Drought emergency in 1990.
- Long-term water supply deficit was also identified.
- City requested proposals to add a water supply.
- Proposal by Ionics, Inc. to design-build-operate was accepted.
- Project was permitted and built under emergency conditions:
  1. RFP: April 1990.
  2. Proposal selection in August 1990.
  3. Environmental Review.
  4. Permits from Coastal Commission, Regional Water Quality Control Board, Corps of Engineers.
  5. Overlapping permitting, design and construction.
  6. Notice to Proceed with Construction: May 1991.
  7. Completed in March 1992 (Ref. 4).

Due to abundant rainfall since 1991, the facility has been on standby since the initial testing period was completed in June 1992 (Ref. 1).

The Santa Barbara City Council decided that the temporary facility would be converted to permanent status for use as a backup during future droughts. The facility also has the potential for use during non-drought periods, which would help meet regional or statewide needs for water by operating under a water exchange agreement. To obtain Charles Meyer Desalination Facility, City of Santa Barbara, California, USA permanent status the facility went through additional environmental review and permitting which was completed in December 1995. The facility has permits to operate as a permanent part of the City's water supply and all equipment is compatible with long term use (Ref. 1). This facility is now fully owned by the City of Santa Barbara (Ref. 4).

### *Permitting and regulatory overview and procedure:*
Current permitted flow rates are based on computer modeling:
"Effluent discharged to the Pacific Ocean shall encounter the seafloor only after the seawater to effluent dilution ratio has increased to the minimum ratio.

The dilution ratio shall be demonstrated by means of a computer model approved by the Executive Officer, employing input variables approved by the Executive Officer.

As estimated by computer modeling, the following table provides: (1) the minimum WWTP discharge flow rate necessary to ensure the combined discharge will remain buoyant and above the seafloor, and; (2) the minimum initial dilution ratio (MIDR) for the combined discharge computed at the minimum POTW discharge flow rate."

*Analyses of plant concentrate:*
Computer model (See above).

*Economic Evaluation:*
The City's facility was built by a private company, Ionics, Inc., under a "take or pay" contract. Over the 5-year contract period, the City, along with the Montecito and Goleta Water Districts, paid off the $34 million construction cost and either paid for water produced or paid to maintain the facility in standby mode.

The cost of desalted water was approximately $1,100 per af including labor, chemicals, power, maintenance, and a sinking fund to replace worn components (Ref. 1).

The plant is currently in "deep storage" and is not operational. Restart costs have been estimated at $10 million or more, with a restart time frame of at least 1 year (Ref 4). [Latest info: $17.7 million (2008 dollars) to reactivate at 3,125 AFY capacity plus $2.5 million for distribution system improvements; $1,470/af operating cost; estimated energy use of 4,615 kWh per af; about 16 months to complete the reactivation. Source: Carollo Engineers, Desalination Rehabilitation Study Final Report, prepared for City of Santa Barbara, March 2009.]

Other institutional constraints on restarting the facility include the perception that operation of the facility would support increased residential and business construction and associated increase in population in the region, and growing concern over the plant's contribution to carbon dioxide levels in the atmosphere (Ref. 2).

*Key project lessons learned:*
Due to the drought emergency, time was a critical factor. Getting power to the site (66kV) was also a significant obstacle (Ref 2).

Being the first large seawater desalination project to go through the California permitting process has its advantages, especially given the emergency motivation. Subsequent project may face additional scrutiny and more stringent conditions.

Taste issues were reported by some customers. It is unclear if these issues were significant, given that some complaints came from customers who were not receiving desalinated water (Ref. 2).

**Pictures**

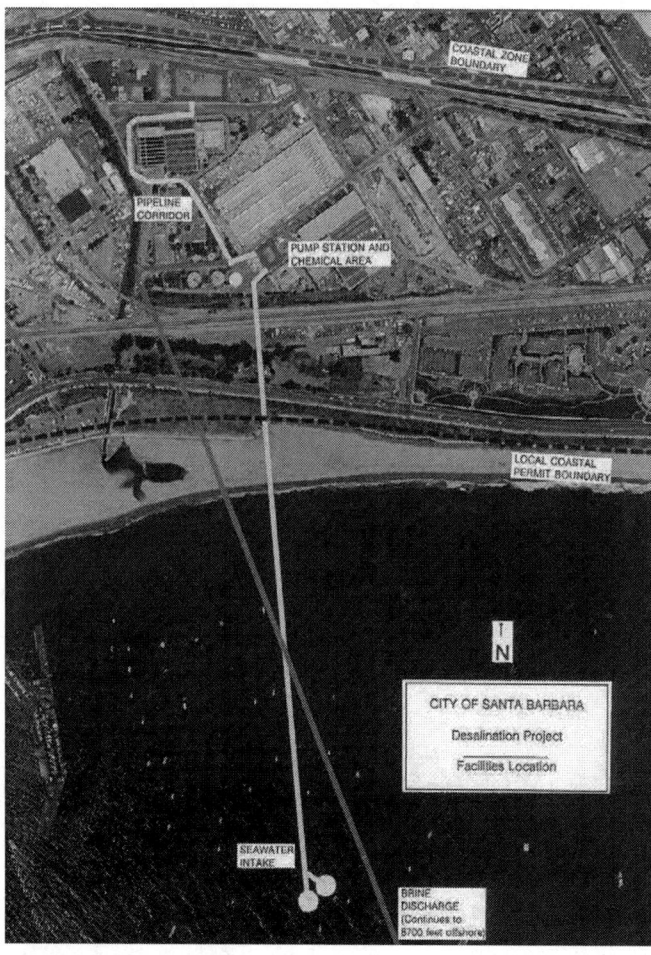

Figure 1. Aerial view of facility, intake line, and a portion of discharge line (Photograph by Bill Ferguson 2004, with permission from County of Santa Barbara Public Works, CA)

Figure 2. Pumping and Chemical Feed (Photograph by Bill Ferguson 2004, with permission from County of Santa Barbara Public Works, CA)

Figure 3. Pre-filtration Tanks with Desal Trailers in Foreground (Photograph by Bill Ferguson 2004, with permission from County of Santa Barbara Public Works, CA)

Figure 4. Modular Capacity: One Membrane Trailer (Photograph by Bill Ferguson 2004, with permission from County of Santa Barbara Public Works, CA)

*Acknowledgement*

Malcolm McEwen developed original case study on the City of Santa Barbara in 2008.

*References*

1. County of Santa Barbara Public Works web page, updated January 22, 2007 www.countyofsb.org/pwd/water/desalination.htm

2. Bjork, Rebecca, Wastewater System Manager, City of Santa Barbara, personal communication, 12/13/2007.

3. State of California, Regional Water Quality Control Board, Central Coast Region, NPDES Permit No. CA0048143 for City of Santa Barbara El Estero Wastewater Treatment Facility.

4. City of Santa Barbara Desalination Presentation, Monterey, California, May 20, 2004, City of Santa Barbara Charles Meyer Desalination Facility, Bill Ferguson, Water Supply Planner.

5. Woodward-Clyde Consultants. Environmental Impact Report for City of Santa Barbara's and Ionics, Incorporated's Temporary Emergency Desalination Project. March 1991.

# Appendix A-2

# Sanitary Sewer and Surface Water Disposal Case Studies

Joe Mullins Reverse Osmosis Water Treatment Facility - Melbourne, FL
Berrin Tansel

Ormond Beach WTP Low Pressure Reverse Osmosis (LPRO) Expansion –
Ormond Beach, FL
Khalil Z. Atasi and Colin Hobbs

Pilot-Research Membrane Treatment of Non-Irrigation Season Flows in the Rio
Grande River – El Paso, TX
Fahy, et al.

## ASCE/EWRI Task Committee
## CONCENTRATE MANAGEMENT IN DESALINATION
### Case Study

Management Approach: Discharge of Concentrate to Brackish Surface Waters

Committee Member(s):     Berrin Tansel

Project Contact(s):     Fred Davis, Superintendent
    Phone: (321) 255-4622, Fax: (321) 255-4636

Project Name: Joe Mullins Reverse Osmosis Water Treatment Facility, Melbourne, Florida

Project Location: 5980 Lake Washington Road Melbourne, FL   32934

Desalination Process: Reverse osmosis of brackish groundwater

WTP Information:

- Rated Capacity: 6.5 MGD
- Max. Concentrate Flow: 6.5 MGD
- Typical Production: 5 MGD
- Typical Concentrate Flow: 1.5 MGD

### *Abstract*

The Joe Mullins Reverse Osmosis Water Treatment Plant was put into operation in 1995 with a 6.5 MGD capacity and a 4.0 million gallon ground water storage tank. Concentrate from the RO plant is discharged into the Eau Gallie River, a Class III marine water body, through an outfall. The outfall is approximately two feet in length and depth, and located approximately 100 feet downstream of the salinity control barrier. The facility currently has the discharge permits by the Florida Department of Environmental Protection (FDEP) and National Pollutant Discharge Elimination System (NPDES). Permits for the treatment facility granted mixing zones for various water quality parameters including dissolved oxygen, total nitrogen, chlorides, specific conductance, pH, gross alpha activity, and combined radium (226+228). Bioassays were conducted for the NPDES permit. Samples were collected for toxicity and algal growth tests. The 96-hour acute definitive toxicity tests were conducted per the permit requirements. The algal growth potential tests showed levels (1.89 mg dry wt/L of *Dunaliella tertiolecta*) which were below the threshold for concern (10 mg dry wt/L of *Dunaliella tertiolecta*). In the near future, the City will need to spend more than $ 19.3 million to expand the treatment capacity of the RO plant to 13.0 MGD.

## *Process Design and Configuration*

Reverse osmosis membrane system treats water obtained from the Lower Floridan Zone of the Floridan Aquifer. Raw water is withdrawn through three 16-inch diameter wells which are approximately 650 to 900 feet deep. The raw water is sent to the Reverse Osmosis Plant for treatment. The filtered groundwater is pumped through a 12" R/O Product Transfer line where filtered waters from both the Joe Mullins Reverse Osmosis Water Treatment Plant (RO WTP) and a Surface WTP are combined and pumped into a baffled contact basin for chlorine disinfection and blending prior to final disinfection and distribution (Kirmeyer, 2004). Addition of fluoride, anhydrous ammonia, and sodium hydroxide (for pH adjustment) occurs within the 45 foot long, 42 inch diameter blended water line connecting the contact basin and baffled reservoir. After post treatment, water is pumped to a secondary unbaffled storage reservoir (4 MG) until distribution.

Concentrate from the current RO plant is discharged into the Eau Gallie River (a Class III marine water body) through an outfall located west of where the river meets the Indian River Lagoon.

The process equipment specification and operational data are provided below (http://www.melbourneflorida.org/watercon/RO%20Details.htm):

| Process Equipment Specifications ||
|---|---|
| **Pre-Filter Data:**<br>Housing Quantity: 2<br>Material: 316 stainless steel; 150 psi code stamped<br>Number of pre-filters per housing: 156<br>Pre-filter Housing: 4' long, 39-1/4"<br>Type of Pre-filter: 5 micron polypropylene | **High Pressure R.O. Pump Data:**<br>Quantity: 2<br>Type: Vertical turbine multi-stage<br>Material: 316 stainless steel<br>Flow Rate: 2270 gpm<br>Motor: 350hp, 1775 rpm |
| **RO Train Data: (M/L 17455)**<br>No. of Trains: 2<br>No. of Pressure Tubes Per Train: 72<br>Staging Array: 48/24<br>Elements per Train: 504<br>Elements per Pressure Tube: 7<br>Material: Thin film composite polyamide<br>Size: 8" diameter by 40" long | **Operating Data:**<br>Number of Trains: 2<br>Inlet Flow Rater per Train: 2125 gpm<br>Product Flow per Train: 1700 gpm<br>Concentrate (brine) Flow per Train: 425 gpm |

| Well Water Analysis: | Permeate Water Analysis: |
|---|---|
| pH: 7.7 | pH: 6.8 |
| Alkalinity: 120 | Alkalinity: 11 |
| Total Hardness: 630 | Total Hardness: 10 |
| Chlorides: 754 | Chlorides: 34 |
| Color: 6 | Color: 1 |
| TDS: 1615 | TDS: 69 |
| Conductivity: 2956 | Conductivity: 127 |

*Project background:*

Melbourne is responsible for providing water to some 150,000 customers. The Joe Mullins Reverse Osmosis Water Treatment Plant (Figure 1 below) was put into operation in 1995 with a 6.5 MGD capacity and a 4.0 million gallon ground water storage tank. To meet future demand, construction of a second phase of the reverse osmosis plant is anticipated in the future, which will provide capacity to accommodate a 100 percent build-out of the current service area (http://www.melbourneflorida.org/pub/pub-pdf/drinkwater.pdf)

*Description of the proposed solution:*

Concentrate from the RO plant is discharged into the Eau Gallie River, a Class III marine water body, through an outfall. The outfall is approximately two feet in length and depth, and located approximately 100 feet downstream of the salinity control barrier.

Assessments conducted by the City of Melbourne and Reiss Environmental, Inc. showed that gross alpha and combined radium were two parameters for which permit limits were being exceeded. Woods Hole Group, Inc. conducted a comprehensive study using a phased approach to evaluate whether a mixing zone could be permitted within the existing water quality regulations (http://www.woodsholegroup.com/project-descriptions/04-125_Melbourne.pdf).

An extensive mathematical modeling effort was conducted for the continued evaluation of the City of Melbourne's reverse osmosis concentrate discharge into the Eau Gallie River. A three-dimensional model of the river was developed using the Environmental Fluid Dynamics Code (EFDC) to simulate the hydrodynamics and particulate transport within the estuarine system. The EFDC model incorporated the parameters for defining the geometry of the system, as well as the conditions at both upstream and downstream boundaries of the Eau Gallie River, the atmospheric conditions, and the concentrate discharge into the model domain. Existing conditions were simulated and the model was calibrated and verified with field data. The model was used to simulate DEP specified design flow conditions to characterize the concentrate dilution and the extent of mixing zones for the parameters of interest (http://www.whgrp.com/bios-long/shultz-bio.pdf).

A consent order granted the facility mixing zones that extend 1,500 feet downstream from the point of discharge and 50 feet upstream from the point of discharge but downstream from the salinity control barriers (DEP, 2004). The mixing zones are monitored to determine if the concentrate disposal causes any environmental damage. The US Environmental Protection Agency and the Florida Department of Environmental Protection are considering Melbourne's applications for permits to allow the by-product to be discharged into an injection well at the D.B. Lee Wastewater Treatment Facility (http://www.melbourneflorida.org/pub/pub-pdf/drinkwater.pdf).

### *Data Collection Procedures:*

Permits for the treatment facility granted mixing zones for various water quality parameters including dissolved oxygen, total nitrogen, chlorides, specific conductance, pH, gross alpha activity, and combined radium (226+228). The parameters are being monitored as defined in the permit.

### *Permitting and regulatory overview and procedure:*

The facility currently has the discharge permits by the Florida Department of Environmental Protection (FDEP) and National Pollutant Discharge Elimination System (NPDES). Bioassays were conducted for the NPDES permit. Samples were collected for toxicity and algal growth tests. The 96-hour acute definitive toxicity tests were conducted per the permit requirements. The algal growth potential tests showed levels (1.89 mg dry wt/L of *Dunaliella tertiolecta*) which were below the threshold for concern (10 mg dry wt/L of *Dunaliella tertiolecta*). The analytical chemistry data showed that concentrate was phosphorus limited. The concentrate samples were not acutely toxic to the fish but acutely toxic to the mysid. The cause of the toxicity was partially due to the high levels of calcium.

### *Analyses of plant concentrate:*

N/A

### *Economic Evaluation:*

In the near future, the City will need to spend more than $ 19.3 million to expand the treatment capacity of the RO plant to 13.0 MGD.
(http://www.ctrlink.com/success/watertreatment.htm)

**Figure 1. Schematic of Joe Mullins Reverse Osmosis Water Treatment Facility, Melbourne, FL (Courtesy of BerrinTansel)**

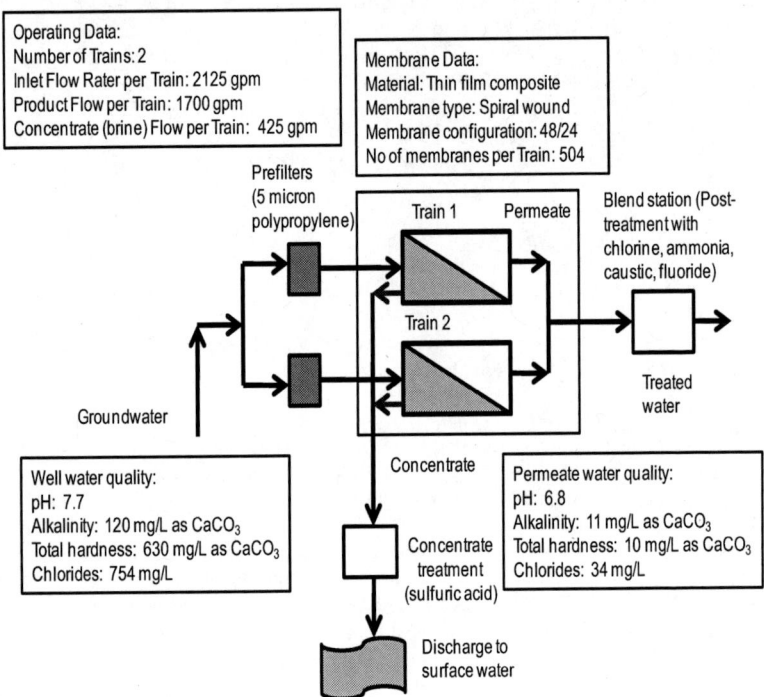

*References*

Kirmeyer, G.J. (2004) Optimizing Chloramine Treatment Second Edition, AWWA Research Foundation U.S.A.
http://www.melbournefl.org/watercon/RO%20Details.htm
http://www.melbourneflorida.org/pub/pub-pdf/drinkwater.pdf
http://www.whgrp.com/bios-long/shultz-bio.pdf
http://www.ctrlink.com/success/watertreatment.htm

DEP, 2004, Individual Environmental Resource Permit and State Lands Approval Technical Staff Report, APPLICATION #: 4-061-75850-2 (State of Florida Department of Environmental Protection/Div of Rec & Parks C/O Bureau of Design & Recreation Services).

## ASCE/EWRI Task Committee
## CONCENTRATE MANAGEMENT IN DESALINATION
### Case Study

Management Approach: Hybrid Zero Liquid Discharge/Surface Water Discharge – Discharge into an existing Public Access Reuse System which provides the ability to reuse all/or a portion of the concentrate for land application or to discharge all/or a portion of the concentrate to a surface water body (Halifax River).

Committee Member(s): Khalil Z. Atasi and Colin Hobbs

Project Contact(s): Colin Hobbs

Project Name: Ormond Beach WTP Low Pressure Reverse Osmosis (LPRO) Expansion

Project Location: Ormond Beach, Florida

Desalination Process: Low Pressure Reverse Osmosis
WTP Information:

- Rated Capacity: 12.0 mgd/4.0 mgd (Maximum permitted capacity of the facility is 12.0 mgd which consists of 8.0 mgd of lime softening capacity and 4.0 mgd of LPRO capacity.)
- Max. Concentrate Flow: 1.0 mgd/0.7 mgd (Maximum permitted concentrate flow is 1.0 mgd, however, the maximum design concentrate flow is 0.7 mgd.)
- Typical Production: Approximately 2.3 mgd (2009)
- Typ. Concentrate Flow: Approximately 0.4 mgd (2009)

*Abstract*

Faced with increasing potable water demands and deteriorating raw water quality, the City of Ormond Beach, Florida (City) retained CDM to design a 4.0-mgd low pressure reverse osmosis (LPRO) expansion to supplement their existing 8.0-mgd lime softening facility. The design of the LPRO expansion was completed in February 2006 and the expansion was placed online in January 2008.

The most innovative aspect of this project focused on concentrate management. Through a collaborative effort, the City, the Florida Department of Environmental Protection (FDEP), and CDM successfully permitted and implemented an innovative and sustainable method of managing a maximum of 1.0-mgd of concentrate. This unique method of concentrate management blends the concentrate with reclaimed water at the City's Wastewater Treatment Plant (WWTP) and allows the City to augment reclaimed water supplies with this previously unutilized resource without consuming WWTP treatment capacity.

Data collected since 2008 demonstrate the success of this innovative and sustainable approach to concentrate management. Since the startup of the LPRO expansion, the City increased reclaimed water supplies by 0.4-mgd and reclaims 100-percent of the concentrate during periods of high reclaimed water demand.

*Process Design and Configuration*

Pretreatment for the LPRO process consists of sulfuric acid addition, antiscalant addition, and cartridge filtration. LPRO treatment is provided by four 1.0 mgd permeate capacity skids operating at 85 percent recovery and an average flux of 14.9 gfd. Post treatment consists of degasification (with subsequent odor control) and free chlorination for primary disinfection. Free chlorinated permeate is blended with filtered and disinfected softened water prior to chloramination, fluoridation, stabilization, corrosion control, and storage.

*Project background:*

Increases in the City's population and service area placed increased demands on the City's existing lime softening WTP and necessitated an increase in the City's potable water treatment capacity. An expansion consisting of additional lime softening units was not feasible due to the gradual intrusion of salt water into the City's groundwater supplies. In order to provide increased potable water treatment capacity and ensure adequate treatment of the City's deteriorating groundwater supplies, the City elected to supplement the existing lime softening WTP with an LPRO expansion, thus necessitating a method of concentrate management. This project commenced in August of 2003 with the initiation of a pilot study and a preliminary design report for the proposed LPRO expansion and concluded when the LPRO expansion was placed online in January of 2008.

*Description of the proposed solution:*

A preliminary permitting meeting with the FDEP, the City, and CDM was held in February of 2004 to present and discuss the potential methods of concentrate management. At the conclusion of the meeting, all parties agreed that blending the concentrate with the effluent at the WWTP was the most favorable alternative. CDM subsequently completed a detailed analysis of this alternative, consisting of a blending analysis to predict potential blended effluent water quality. The results of this analysis were submitted to the FDEP along with the required permit forms to modify the WWTP in October of 2004. In November of 2004, CDM received one request for additional information (RAI) from the FDEP related to the significance of saline irrigation supplies, specifically pertaining to their effects on vegetation and soils. The FDEP received and reviewed CDMs responses and the FDEP granted the requested modifications to the WWTP in March of 2005.

Please note the project originally called for a 2.0 mgd, expandable to 4.0 mgd, LPRO expansion. During the construction of the LPRO expansion, the City elected to install all 4.0 mgd of LPRO capacity. As such, it was necessary to re-permit the WWTP. Repermitting efforts commenced in March of 2007 and concluded in October of 2007. During the re-permitting process CDM received two RAIs from the FDEP.

*Data Collection Procedures:*

Data utilized throughout this project included: projected LPRO concentrate water quantity and quality, historical wastewater quantity and quality (including quantities discharged to the Halifax River and to the Public Access Reuse System), historical groundwater quality, a survey of all existing wells within the Public Access Reuse System service area, and an identification of soil and turfgrass types within the Public Access Reuse System service area. The membrane manufacturer provided the projected concentrate water quantity and quality based upon the design operating conditions; the City provided historical wastewater quantity and quality, historical groundwater quality, and the survey of all existing wells; and a review of relevant literature allowed for the identification of soil and turfgrass types.

*Permitting and regulatory overview and procedure:*

Prior to the preliminary permitting, six concentrate management alternatives were identified and preliminarily evaluated. These alternatives were presented and discussed at the preliminary permitting meeting between all key parties. At the conclusion of the preliminary permitting meeting, the FDEP, the City, and CDM identified one alternative as the most favorable and all subsequent efforts were focused on this alternative. This meeting provided an open and collaborative forum for all parties to discuss the concentrate management alternatives, voice opinions and concerns, and take an active role in the decision making process. Furthermore, this meeting set the tone for all future efforts related to the concentrate management process and greatly facilitated the permitting process. Subsequent meetings were scheduled to present and discuss the responses to the FDEPs RAIs.

*Analyses of plant concentrate:*

The FDEP granted the requested WWTP permit modification and included the following specific conditions:

- A maximum of 1.0 mgd of demineralization concentrate may be blended with the reclaimed water
- The maximum reclaimed water:demineralization concentrate ratio must not exceed 2.5:1.0
- The blended reclaimed water must be monitored for pH, specific conductance, fluoride, sodium adsorption ratio, sodium, magnesium, calcium, chloride, total suspended solids, and total dissolved solids
- The groundwater monitoring plan must be updated to include quarterly monitoring for sodium, sulfate, and radium 226 + 228

*Economic Evaluation:*

Minimal capital costs were associated with the implementation of this concentrate management alternative due to the proximity of the WTP to the WWTP and the ability to slip-line an existing and abandoned forcemain for a majority of the concentrate pipeline length. Additionally, the City was able to reclaim and reuse a previously unutilized resource.

*Key project lessons learned:*

1. Approach the permitting authorities as early as possible in the project so all parties are involved in key decisions.
2. Develop and maintain open lines of communication with the permitting authorities throughout the project.
3. Plan for minimum daily flow events or provide some method of flow equalization or storage.

**Pictures**

**Figure 1. Ormond Beach Project Location**
Source: Hobbs et al. 2009. Reproduced with permission from American Water Works Association.

**Figure 2. Ormond Beach WTP Vertical Turbine Membrane Feed Pumps**
Source: Poster session in 2008 by Colin Hobbs. Reproduced with permission from American Water Works Association.

**Figure 3. Ormond Beach WTP LPRO Membrane Skids**
Source: Poster session in 2008 by Colin Hobbs. Reproduced with permission from American Water Works Association.

**Figure 4. Ormond Beach WTP Concentrate Blending**
Source: Hobbs et al. 2009. Reproduced with permission from American Water Works Association.

*Acknowledgements:*
The authors would like to acknowledge the valuable assistance and information provided by the City throughout this project. Furthermore, we would like to acknowledge input and guidance provided by the FDEP throughout the permitting process.

*References:*

Hobbs, C., Rasmussen, D., and Noble, J., "A Sustainable Approach to Low Pressure Reverse Osmosis Concentrate Management: Reclaiming and Reusing a Valuable Resource," Proceedings from the 2009 AWWA Annual Conference and Exposition, San Diego, CA, June 14-18, 2009.

Unruh, J. B. and Elliott, M. L., Best Management Practices for Florida Golf Courses, University of Florida Institute of Food and Agricultural Sciences, Gainesville, FL, 1999.

## ASCE/EWRI Task Committee
## CONCENTRATE MANAGEMENT IN DESALINATION
### Case Study

Discharge of Concentrate to Sanitary Sewer or Surface Waters

Committee Member(s):   Michael Fahy, EPWU

Project Contact(s):   Michael Fahy, John Balliew, EPWU
Dr. Anthony J. Tarquin, UTEP

Project Name: Pilot-Research Membrane Treatment of Non-Irrigation Season Flows in the Rio Grande River

Project Location: El Paso, TX

Desalination Process: Nanofiltration and Reverse Osmosis (Pilot Plant)

**WTP Information:**

- Rated Capacity: 3 gallons per minute (gpm) (Pilot-Scale Operation)
- Max. Concentrate Flow: 1.0 gpm
- Typical Production:   3.0 gpm
- Typ. Concentrate Flow:   0.7 gpm

*Abstract*

Using impaired waters to meet increasing municipal demands has become nearly mandatory in many locations. This pilot scale project was undertaken to determine the feasibility of using brackish irrigation return flows (that also contain a small percentage of treated wastewater) to satisfy part of El Paso's municipal water demand during the non-irrigation season.

The first step was to determine if a sufficient quantity of water is available during the non-irrigation season to provide 10 MGD of potable water. Historical data obtained from the United States Geological Survey for the years 1936 through 2004 revealed that a flow of at least 20 MGD is available 89% of the time.

A simple microfiltration system (MF) was installed ahead of the reverse osmosis (RO) and nanofiltration (NF) membranes for suspended solids removal. The desalting results showed that both the RO and NF systems performed very well, with salt rejections in the 98% range. The lowest cost blended water was $0.76 per thousand gallons, produced in the NF unit when operating at 170 psi with a flux of 25 gallons per square foot per day. This cost includes disposal of concentrate through re-blending with available irrigation return flows for possible irrigation of onions, the major winter crop in El Paso County, Texas.

## *Process Design and Configuration*

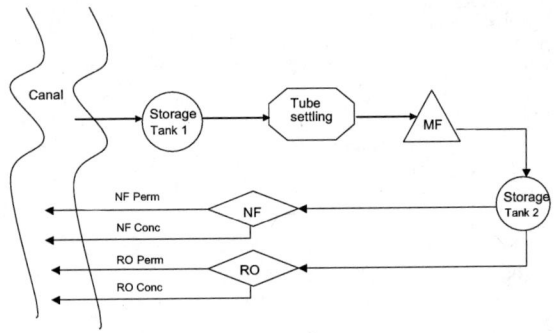

**Figure 1. Schematic of Pilot Treatment System**

## *Project background:*

The El Paso Water Utilities (EWPU) receives as much as 85% of its potable water supply from the Rio Grande during the irrigation season of the U. S. Bureau of Reclamation's Rio Grande Project. The irrigation season normally runs from March through September during a typical water supply year. The raw water quality is acceptable for treatment to potable water standards using conventional treatment methods during this period. However, when raw water releases from Elephant Butte and Caballo Reservoirs are curtailed at the end of the regular, irrigated growing season, the river water consists primarily of return flows from agricultural drains. These simple, incised drains were excavated to remove water from the root zone, thereby relieving the soil of excess water and improving crop production.

EPWU must curtail diversions from the river during the non-irrigation season (approximately October through February) because the total dissolved solids (TDS) content of Rio Grande water during that time generally exceeds 1000 mg/L. This represents a considerable portion of the year during which surface water is essentially unusable as a potable supply due to high TDS. This project was undertaken to determine the feasibility of treating river water after the irrigation season ends (i.e. irrigation return flows) to meet potable water quality standards using membrane technology, thereby utilizing an available water resource that is currently not put to full beneficial use.

## *Description of the proposed solution:*

The first step in the project involved assessing the availability, reliability, and quantity of winter return flows in the Rio Grande. If the water available during the non-irrigation season is insufficient from a quantitative standpoint, any qualitative evaluation would be irrelevant. Therefore, historical data were obtained from the United States Geological Survey (USGS) for the years 1936 through 2004 at the Courchesne Bridge and analyzed for average, median, and low flows during the non-irrigation season. Once the quantity of water available was determined to be adequate, the same database was used to obtain water quality parameters as a function of flow. Treatability studies were then conducted using pilot plants containing different types of membranes as discussed below. This part of the study also generated design and cost information for possible full-scale implementation of the project.

Three different pilot units were set up at EPWU's Canal Street surface water treatment plant site to collect data for this study. The MF unit, which was used for suspended solids removal, consisted of four, 4-inch (10 cm) diameter microfiltration membrane housings 40" (100 cm) long connected in parallel and mounted horizontally on a stand, one above the other. The other two pilot units were single-membrane GE-Osmonics units, one equipped with a nanofiltration membrane and the other with an RO desal membrane. Ball valves at the end of each housing allowed for any of the membranes to be isolated for cleaning or other servicing. The antiscalant Pretreat Plus Y2K from King Lee Technologies was added at 2 ppm during all testing.

In addition to the treatment units themselves, there were two storage tanks included in the treatment system. All pilot equipment (i.e. everything except the two storage tanks) was housed in a 12'x 24' (3.7 x 7.3 m) building located close to the American Canal. A schematic of the pilot treatment system is shown in Figure 1. For a full-production scale plant, permeate water would be blended back with river return flows treated in the conventional treatment plant and pumped into the potable water distribution system.

### Data Collection Procedures:

Data collection activities for this project spanned parts of the 2006 thru 2008 irrigation and non-irrigation seasons. The pilot plants were operated on an intermittent basis, ranging from a few hours at a time to several weeks continuously, depending on the task at hand. The water quality parameters that were measured also varied according to what was being investigated at the time. Some of the water quality parameters were measured at the site, but most were measured in the lab. Besides the operating parameters that obviously had to be recorded at the site (such as temperatures, pressures, flows, etc.), some of the conductivity, pH, and turbidity readings were also made at the site. Conductivity and temperature readings were made with an Oakton model Con 110 combination pH, conductivity, TDS meter. Turbidity readings were made with a HACH model 2100N Turbidimeter. The more routine analyses such as alkalinity, chlorides, hardness, etc. were done at The University of Texas at El Paso using HACH standard procedures. The more robust sample analyses were done at the International Water Quality Laboratory of EPWU, including analyses for anions and cations, total organic carbon, silica, etc.

### Permitting and regulatory overview and procedure:

Research nursery testing of the RO and Nano concentrates for irrigation of commercial onions indicates that the concentrate may be successfully blended back with raw winter return flow waters and used for irrigation. Since the concentrate can easily be discharged back into the raw water source canal at this plant location, and thereby blended back with additional raw winter return flow waters for downstream winter irrigation, there should theoretically be no additional cost for concentrate disposal.

### Analyses of plant concentrate:

The RO membrane was a GE - RO desal membrane and the nano membrane was a Filmtec nano NF90-4040 membrane. Table 1 shows some of the chemical parameters for the permeate and concentrate from each membrane. As the data show, both membranes performed very well from the standpoint of salt rejection, with the nano membrane performing as well as the RO membrane. (There is a wide spectrum of the performance in nanofiltration membranes).

## Table 1. Permeate & Concentrate Quality from RO and Nano Membranes

| Parameter | Tank 2 (Feed) | RO Perm | RO Conc | Nano Perm | NanoConc |
|---|---|---|---|---|---|
| Calcium | 95.1 | <10 | 364 | <10 | 380 |
| Magnesium | 25.8 | <0.5 | 97 | <0.5 | 102 |
| Potassium | 12.3 | <2 | 44.1 | <2 | 46.5 |
| Sodium | 396 | 25.9 | 1480 | 16.1 | 1560 |
| Bromide | <0.05 | <0.05 | <0.50 | <0.05 | <0.50 |
| Chloride | 341 | 30.7 | 1250 | 16.89 | 1360 |
| Fluoride | 0.76 | <0.1 | 2.33 | <0.1 | 2.43 |
| Nitrate-N | 2.68 | 0.97 | 6.16 | 0.66 | 6.94 |
| Nitrite-N | 0.25 | 0.07 | 0.56 | <0.05 | 0.62 |
| Ortho-P | 0.3 | <0.05 | 0.82 | <0.05 | 0.94 |
| Sulfate | 427 | 4.8 | 1640 | 1.26 | 1760 |
| Alkalinity | 226.8 | 11.7 | 771 | <5 | 880.3 |
| Conductivity | 2360 | 150 | 7500 | 79.5 | 8000 |
| TDS | 1450 | 62 | 5410 | 58 | 5790 |
| pH | 8.2 | 7.1 | 8.2 | 6.9 | 8.2 |
| Silica | 20.3 | <5.0 | 77.4 | <5.0 | 83.9 |

Note: All units mg/L except Cond ($\mu$S/cm ) and pH

Source: Data from "Membrane Treatment of Impaired Irrigation Return and Other Flows for Creating New Sources of High Quality Water." Copyright 2010. Water Research Foundation. ALL RIGHTS RESERVED.

The flux in the RO unit was 12.7 gallons per square foot per day (gfd) at a recovery rate of 73%, with the values in the nano unit at 23.5 gfd and 75%, respectively. The results are for treatment that occurred in the month of February during a period of prolonged, multi-year, regional drought. The raw water conductivity of 2440 $\mu$S/cm is approximately the 93$^{rd}$ percentile of non-irrigation season conductivities, so these results can be considered as nearly worst-case conditions. Even so, the TDS of both permeates is so low that a considerable amount of blending can be done. The silica concentration of 84 mg/L in the nano concentrate is well below the concentration that would be of concern with respect to membrane fouling, even if an antiscalant weren't used. Thus, in general, these results appear to be very favorable from the standpoint of system performance.

### *Economic Evaluation:*

Since both the RO and nano systems performed well from the standpoint of salt rejection, an economic analysis was conducted to determine the optimum operating conditions for each system based on the cost of the produced water. Data were collected over a full range of permeate flows and pressures for both systems by changing the variable frequency drive (VFD) settings from 25% to 75% and adjusting the recoveries from 35% to 90%. The data were collected in three different time periods: November, 2006, and January and February of 2007. These data were then used to project the capital and operating costs of a full scale system so that the unit cost of water could be determined for each condition.

The values used in the cost calculations are shown in Table 2. Note that the costs are based on a permeate water volume of 5 MGD and a final product water quality of 750 mg/L TDS. Blending would be approximately 1:1, using pretreated river water, so the total net river water removal volume would be about 10 MGD. Membrane concentrate is assumed to be discharged back into the raw water canal for winter crop irrigation, or for eventual return to the river. Personnel costs for the 5-month operating season are estimated at $100,000. Contingency costs of $100,000 per season are also included in the calculations. The unit costs are plotted as a function of pressure in Figures 2 and 3 for the RO and nanofiltration units, respectively.

**Table 2. Values Used in Financial Calculations**

| Item | Value |
|---|---|
| Permeate volume, MGD | 5 |
| River TDS, mg/L | 1400 |
| Final product water TDS | 750 |
| Raw water cost, $/acre-ft | $16.00 |
| Pretreatment cost (Canal plant), $/1000 gal | $0.15 |
| Interest rate, % | 5% |
| Equipment life, yrs | 20 |
| Recovery of pretreated water,% | 96% |
| RO capital cost, $/ft$^2$ membrane | $8.64 |
| Power cost, $/kw-hr | $0.07 |
| Pump & Motor efficiency, % | 75% |
| Building Cost, $ | $500,000 |
| Building Life, yrs | 30 |
| Membrane cost, $/80 sq ft | $300 |
| Membrane life, yrs | 3 |
| Antiscalant, $/9 lb gal | $11.00 |
| Antiscalant dosage, mg/L | 2.0 |
| Personnel, $/yr | $100,000 |
| Contingencies, $/yr | $100,000 |

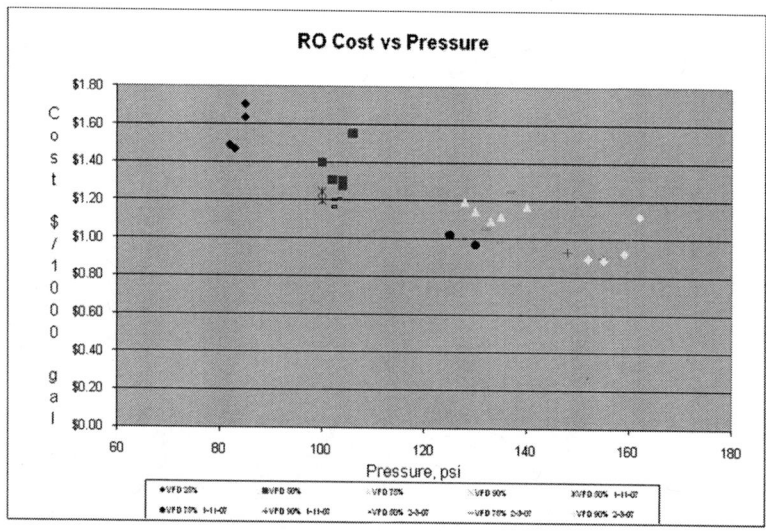

**Figure 2. RO Unit Water Costs versus Pressure**
Source: Data from "Membrane Treatment of Impaired Irrigation Return and Other Flows for Creating New Sources of High Quality Water." Copyright 2010. Water Research Foundation. ALL RIGHTS RESERVED.

The results show that the unit cost of the final product water (i.e. blended water) generally decreases as the pressure and recovery increase in both the RO and nano units. The pressure effect is more pronounced for the RO unit than for the nano. For the RO unit, the lowest costs are generally associated with recoveries around 75% and pressures in area of 150 psi. The lowest costs are in the $0.89 to $0.96 per thousand gallon range for the 2007 data, which were collected under conditions that were more normal than in 2006. The 2006 data gave higher unit costs because the water was a little saltier at that time. However, operating at such high pressures results in high fluxes, which may not be sustainable for a continuously operating membrane plant. Even so, the cost for desalting Rio Grande irrigation return flows in a reverse osmosis system would still be about the same as water produced in the conventional surface water treatment plants during the normal irrigation season (because of the much lower cost of the raw water).

For the nano unit, the costs are lower than for RO over a rather broad pressure range, but the lowest unit costs are generally associated with pressures above 170 psi and recoveries in the 60% to 80% range. Since the fluxes above 200 psi tended to exceed 30 gfd, a pressure of around 170 psi would be possible for the final design, if it were proven to be sustainable. However, an operating pressure of approximately 110 psi would likely be preferable for final design, which would result in a unit water cost of approximately $1.00 per thousand gallons, based on the 2007 data.

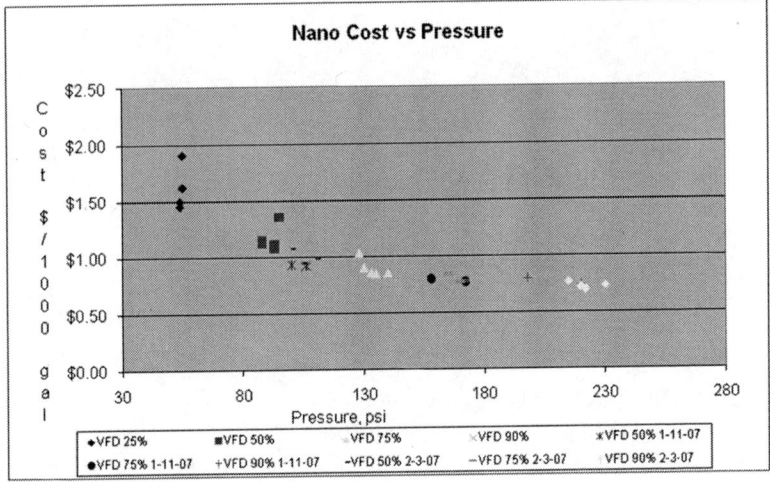

**Figure 3. Nano Unit Water Costs versus Pressure**
Source: Data from "Membrane Treatment of Impaired Irrigation Return and Other Flows for Creating New Sources of High Quality Water." Copyright 2010. Water Research Foundation. ALL RIGHTS RESERVED.

*Key project lessons learned:*

Based on the results of this investigation, the following conclusions can be made with reasonable certainty:

1. From a technical point of view, desalting of Rio Grande irrigation return flows during the non-irrigation season appears to be quite feasible for producing at least 10 MGD of drinking water for EPWU customers.

2. A nanofiltration system using a Filmtec nano NF90-4040 nanofiltration membrane will produce drinking water having a slightly lower cost than a reverse osmosis system. The unit cost of water produced in such a system is projected to be about $1.00 per thousand gallons.

3. Several nanofiltration full-scale system designs would produce water in the $0.76 to $0.79 per thousand gallon range, but they would not be sustainable due to the high flux generated. The design that would seem to produce the least amount of stress in the system would operate at 110 psi with a recovery of about 68% and a flux of 13 gfd.

4. Both of the membranes tested in this study (RO and nano) produced very high quality water (TDS concentrations generally less than 100 mg/L). This will allow for a high blend ratio of 1:1, which results in low-cost product water, rendering the project very attractive from an economic point of view.

5. Winter irrigation of commercial onion crops is possible using RO or Nano concentrate during periods when an adequate supply of raw winter return flows is available for blending prior to application. Careful management would be required for germination and crop establishment.

**Pictures**

Photo 1. View of Pilot Plant Pre-treatment including Microfiltration (Photograph by Michael Fahy 2006, with permission from El Paso Water Utilities, El Paso, TX)

**Photo 2. Dr. Tarquin standing in front of Pilot Nanofiltration and RO Membranes (Photograph by Michael Fahy 2006, with permission from El Paso Water Utilities, El Paso, TX)**

*Acknowledgments*

Funding and support for this project was provided by the following organizations: American Water Works Association Research Foundation, U.S. Bureau of Reclamation, El Paso Water Utilities, the U.S. EPA through the Southwest Center for Environmental Research and Policy, Texas A&M AgriLife Research Center at El Paso, U.S.D.A.-CSREES Rio Grande Basin Initiative, and the University of Texas at El Paso.

*Reference*

Riley, R. T., "Treatment of Non-Irrigation Water at the Canal Street Water Treatment Plant", Master of Science Thesis, Department of Civil Engineering, August 2005, University of Texas at El Paso.

# Appendix A-3
# Deep Well Injection, Land Disposal, and Evaporation Ponds

EPWU Deep Well Injection Plant – El Paso, TX – Mike Fahy & Ken Mercer

North Collier County Regional Water Treatment Plant - Naples, FL – Berrin Tansel

Central Plantation Water Treatment Plant - Plantation, FL – Berrin Tansel

Dalby Stage 2 Desalination Plant – Queensland, Australia – James Jensen

## ASCE/EWRI Task Committee
## CONCENTRATE MANAGEMENT IN DESALINATION
### Case Study

Disposal Method: Deep Well Injection

Committee Member(s): Kenneth Mercer, Mike Fahy

Project Contact(s): Mike Fahy, Scott Reinert, EPWU

Project Name: Kay Bailey Hutchison (KBH) Desalination Plant

Project Location: El Paso, TX

WTP Information:

- Rated Capacity: 27.5 MGD
- Max. Concentrate Flow: 3 MGD
- Typical Production: 3.3 MGD
- Typical Concentrate Flow: 0.5 MGD

### *Process Design and Configuration:*

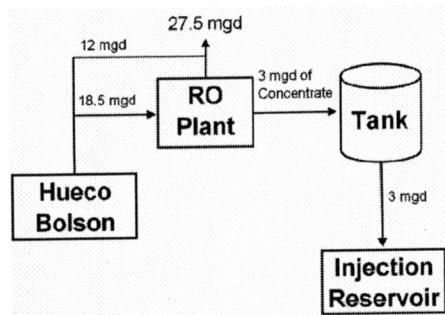

There are 36 production wells that are connected to the desalination plant. These wells are a combination of existing wells (4), rehabilitated wells (3), and new wells (29), and are connected to the plant by 88,700 ft of new collector lines. The location of the wells is designed to intercept brackish groundwater. The drawdown trough that will be created will prevent further migration of brackish groundwater into areas that have historically pumped fresh groundwater.

The total dissolved solids (TDS) content of the raw feed water ranges from a low of approximately 1,500 to 2,000 ppm at the current, typical production rate when only a few wells are in operation, to more than 3,000 ppm when the plant is operated at full capacity.

The reverse osmosis plant has five skids of membranes manufactured by Hydranautics. Each skid is capable of flow rates between 1.70 and 3.64 mgd. Expected recovery is between 70% and 82.5%, and expected salt rejection is up to 93%. The plant inflow will be treated with an antiscalant to control mineral precipitation (notably silica) and hydrochloric acid to adjust pH. The finished water will be treated with caustic soda to raise the pH, a corrosion inhibitor, and disinfected with sodium hypochlorite. The concentrate can be treated with hydrochloric acid to adjust the pH prior to disposal, if needed.

There is space available within the treatment building for installation of additional membranes if raw water quality were to deteriorate, thus reducing the effective recovery of the facility. The engineer's design life for the plant is approximately 45 years. EPWU's plan is to replace each membrane every 5 years due to loss in efficiency. However, each membrane could, in reality, last as long as 15 years.

### *Concentrate Management Alternatives Considered:*

Passive evaporation, enhanced evaporation, and deep well injection

### *Factors that influenced management strategy selection including applicable regulatory agencies (rank if possible):*

Cost, land requirements, subsurface conditions

### *Project background:*

El Paso has relied on the Hueco Bolson as a major water supply source since 1903. As a result of high pumping, groundwater levels in the Hueco have declined and brackish groundwater has intruded into areas that historically yielded fresh groundwater. El Paso Water Utilities (EPWU) began reducing its Hueco pumping in 1989. This action was made possible by a variety of water management initiatives including increased water conservation, increased surface water diversions, and increased reclaimed water use. The reduction in pumping has resulted in stabilized groundwater levels in many areas. However, brackish groundwater intrusion remains an issue. A 27.5 mgd desalination plant has recently been completed that will result in reductions in brackish groundwater intrusion, and allow EPWU to better utilize its fresh groundwater wells during droughts. The location of the pumping wells will provide an opportunity to intercept the brackish groundwater before it intrudes into historically fresh groundwater areas.

Because of the attractive costs, a detailed investigation of the deep well disposal option was completed from 2002 to 2004, and consisted of geologic investigations, test drilling, geophysical studies, preliminary modeling, and culminated in the construction and testing of a pilot well. The geophysical investigation consisted of gathering existing gravity measurements in the area and collecting new gravity measurements. All data were interpreted with respect to formation depths obtained from the test hole drilling program. The resulting subsurface geologic model provided the basic framework on which a preliminary numerical flow model of the area was constructed.

The objective of the preliminary numerical flow model was to investigate the potential range of relative hydraulic conductivity conditions in the various rock units. In addition, the role of faults in the area was evaluated (i.e. conduits or barriers to flow). The results included estimates of the potential build-up of water levels and area of concentrate migration under a wide range of geologic and operational scenarios.

### *Description of the proposed or implemented solution:*

Kay Bailey Hutchison Desalination Plant uses three deep injection wells located over 20 miles from the plant for the disposal of up to 3 million gallons per day (mgd) of RO concentrate. The three injection wells are between 3,700 and 4,000 ft deep, and are completed in the Silurian Fusselman Formation, a fractured dolomite, and the underlying Montoya Formation, a fractured limestone. The Surface Injection Facilities consist of yard piping, a 300,000 gallon storage tank at each site, a solar power system (with generator backup), and various instrumentation and controls to manage the injection and collect performance data. Figure 1 of this report is a photograph of the drilling rig in operation during construction of an injection well. Figure 2 is a photograph of a wellhead for one of the completed injection wells.

### *Permitting and regulatory overview and procedure:*

The results of the studies and the testing of the pilot well were used in 2004 and 2005 to prepare an application to the Texas Commission on Environmental Quality (TCEQ) for a Class V Authorization to inject concentrate into the Fusselman and Montoya formations of Silurian and Ordovician age. The key features of the application were that the proposed injection wells were in a remote location (i.e. no other production or injection wells in the area), the expected concentrate had a lower total dissolved solids concentration than the formation water (about 8,000 mg/l), and the injection would be by "gravity" (i.e. no injection by pumping). The application acknowledged the limitations of a single short-term test, and acknowledged the uncertainty of the faults as barriers and/or conduits to flow. The authorization was obtained on July 13, 2005.

### *Guidelines for analysis of plant concentrate:*

During the development of the TCEQ application and after the authorization was obtained, several studies were completed related to the potential for mineral precipitation in the well and formation. Of notable concern were calcite, barite and silica. Jar testing of the concentrate, formation water and formation was completed, as was preliminary geochemical modeling. Mitigation strategies were identified (e.g. lowering the pH of the concentrate), and a plan for initial operation was developed to further test the potential for mineral precipitation during initial operation.

The second and third injection wells were constructed in 2006 and the beginning of 2007. These wells were also completed to Class I standards. The wells are 3,720 ft and 4,030 ft deep, and both include open-hole completions in the injection zone (below 2,900 ft). During test pumping of each of the wells, drawdown data were collected in the two non-pumping wells to provide estimates of aquifer transmissivity. The data obtained from the construction of these wells were combined with previous data to update the subsurface geologic model.

At the completion of testing, each well was video logged to assess the nature and size of the fractures in the injection zone. Numerous fractures over the entire thickness of the injection zone were observed, many of which were nearly an inch wide. The number and size of the fractures, coupled with the open-hole completion reduced concerns regarding the potential for mineral precipitation.

Initial testing of the wells began in May 2007 and initially involved injecting fresh water routed from the desalination plant in order to develop well performance data without concern of mineral precipitation. At the beginning of plant operations, the concentrate was diluted, thus continuing baseline data collection. After sufficient baseline data were collected, the dilution of the concentrate was reduced and finally eliminated, with no observable change in well performance (i.e. injection rate and groundwater level build-up). During these tests, the concentrate received no treatment (e.g. no pH adjustment). Although the tests were short-term and the initial operation has been only a few months, it appears that mineral precipitation is not significant with respect to well performance. Monitoring efforts continue in order to assess the potential for mineral precipitation and interpret long-term reservoir performance.

***Key project constraints:***

The four key features of the application for implementation of the injection wells were that the proposed injection wells were in a remote location, the expected concentrate had a lower total dissolved solids concentration than the formation water (6,000 mg/l vs. 7,000 mg/l), the injection would be by "gravity" (i.e. no injection by pumping) which would reduce or eliminate the potential to fracture the reservoir further, and the maximum rise in groundwater level would be about 160 ft over a ten year period (thus final depth to water would be about 329 ft below ground surface). The application acknowledged the limitation of a single short-term test, and acknowledged the uncertainty of the faults as barriers and/or conduits to flow.

The current Class V injection well authorization prohibits injecting water that does not meet primary drinking water standards, even if the formation water exceeds the primary drinking water standard for that particular parameter. Native Fusselman-Montoya-El Paso Group water samples demonstrate that the water quality does not meet national and state primary drinking water standards for arsenic, gross alpha (less Ra and U), nitrite, and radium. In addition, the formation water is brackish with a TDS of over 8,000 mg/L.

Under current operations, the chemical composition of the desalination concentrate (injectate) has a TDS less than 6,000 mg/L. Thus, the concentrate has an overall higher quality than the native Fusselman-Montoya-El Paso Group water. The only parameters of the concentrate that do not meet primary drinking water standards are arsenic and gross alpha (less Ra and U). As noted above, the native Fusselman-Montoya-El Paso Group formation water contains arsenic and gross alpha that already do not meet primary drinking water standards.

Currently, the concentrate is being diluted in order to meet the requirements of this authorization (i.e., arsenic and gross alpha concentrations below primary drinking water standards). While the plant is currently generating only 700 gallons per minute (gpm) of concentrate, EPWU recognizes that as water demand increases over the years, the volume of concentrate will also increase, raising the question of how to address the primary drinking water standard issue.

The most viable option in dealing with injecting concentrate that does not meet primary drinking water standards for one or more parameters is an "aquifer exemption." The U.S. Environmental Protection Agency (EPA), TCEQ, and New Mexico Environment Department (NMED) can jointly approve an aquifer exemption by finding that this use (injecting concentrate) in a USDW aquifer may be more important than or otherwise take precedence over, the use of the aquifer as a potential source of water supply for human consumption.

Aquifer exemptions require modifications to State Underground Injection Control (UIC) Programs, including public notice and participation. The exemptions are granted by TCEQ and NMED with concurrence from the EPA in accordance with 40

CFR Parts 144-146, 30 TAC Chapter 331, and 20.6.2.5103 NMAC. The process includes submittal of an application package to TCEQ and NMED for review. Once the TCEQ and NMED reviews and tentatively approves an aquifer exemption request, the request is sent to EPA for approval.

Aquifer exemptions may be granted under EPA 40 CFR §146.4, TCEQ 30 TAC 331.13, and NMED 20.6.2.5103, if:

> a) Aquifer is not a source of drinking water and will not serve as a source of drinking water in the future because it- has a TDS level above 3,000 mg/L and less than 10,000 mg/L, and is not reasonably expected to supply a public water system; and

> b) Is too deep or too remote which makes recovery of water for drinking water purposes economically or technically impractical.

EPWU's position is that both of these conditions apply to the KBH site. Due to a revision of injection scenarios in the groundwater model, the plume of arsenic which exceeds the primary standard no longer extends into the State of New Mexico. Therefore, EPWU will withdraw its request from the State of New Mexico for an aquifer exemption.

The following is a breakdown of the recent milestones during the Exempt Aquifer Application Process:

- EPWU Applied for Aquifer Exemption from Anti-Degradation Requirement: Discussions with TCEQ began in December 2007
- August 20, 2008- Exempt Aquifer Application submitted to TCEQ.
- September 3, 2008- Exempt Aquifer submitted to NMED.
- March 31, 2009- EPWU received written comments from NMED.
- April 28, 2009- LBG/EPWU met with TCEQ to discuss Exempt Aquifer Application.
- June 4, 2009-EPWU received written comments from TCEQ.
- June 9, 2009-EPWU collected water samples from selected JDF wells and the plant as a response to TCEQ comments.
- June 30, 2009-EPWU met with NMED on June 30, 2009 to discuss review comments.
- September 2009-EPWU will send response to comments with page revisions to the original application to TCEQ.

### *Economics:*

The total, or cumulative, capital costs incurred for the 21 individual construction contracts for the project equaled $91 million, and is comprised of the following:

a) Production Wells and Collectors - $ 32 Million
b) Plant and Near-Plant Pipes - $ 40 Million
c) Concentrate Disposal - $ 19 Million

The original amortized (capital plus O&M) cost for the project was estimated to be $534/acre foot ($/AF), based on a production rate equal to 80% of design capacity and a 5% discount rate. However, the actual O&M cost alone during the 2008/2009 fiscal year equaled approximately $571/AF, or $1.75/1000 gallons, due to the lower than expected production rate. Also, the actual electrical rate during 2008/2009 equaled 8.5 cents per Kilowatt-Hour (KWH), rather than the estimated rate of 7 cents/KWH originally anticipated.

## *Key project lessons learned:*

In December of 2005, EPWU convened a QA/QC panel of consultants not otherwise associated with the project to review the following aspects of concentrate disposal:

- Wet well design
- Pipeline design and hydraulics
- Potential for mineral precipitation
- Surface injection facilities
- Injection well design
- Injection reservoir

The panel raised concerns regarding the reservoir capacity and that the initial tests were not adequate to fully characterize the reservoir. However, the panel also noted that additional testing and monitoring were planned. The panel also raised concerns regarding the potential for mineral precipitation. They noted that the antiscalant that will be added to the raw water will inhibit precipitation in the wet well, concentrate pipeline, storage tank and well bore. However, it is expected that the antiscalant will adsorb onto the formation and mineral precipitation could occur in the formation. The potential for precipitation, however, will be dependent on the kinetics of the reaction, which is a function of the relative mix of concentrate and formation water, the temperature and the pH. The significance of the precipitation will be dependent on the size of the fractures and the velocity of flow away from the well bore.

Geochemical Technologies Corporation (GTC) conducted a study to evaluate the potential for minerals to precipitate from solution in the pipeline during transport, in wells during injection, or after injection in the Fusselman Formation.

Based on analysis of the concentrate that is to be injected, it was determined that calcite, barite, aluminum and iron hydroxide, and silica are supersaturated and have the potential to form a precipitate in the formation, pipe or the well bore. The injection flux may be sufficient to carry the precipitates out into the formation further away from the well bore region. The nature and extent of fracturing in the limestone will determine the degree to which precipitate formation limits the hydraulic properties of the aquifer to receive concentrate.

Potential mitigation measures for mineral precipitation currently identified include pH adjustment (acid pumps can introduce HCl in the concentrate stream as it enters the concentrate pipeline), dilution of concentrate with brackish groundwater, and lime treatment of the concentrate. Monitoring changes in pressure buildup should reveal if any precipitation has initiated. In the event that changes in pressure buildup are observed, the plant is equipped with the capacity to add acid to lower the pH of the concentrate and to slow or prevent precipitate buildup. Acid is not now being added to the concentrate.

**Pictures**

**Figure 1. Well Construction (Photograph by Eric Bangs 2007, with permission from El Paso Water Utilities, El Paso, TX)**

**Figure 2. Well-head (Photograph by Eric Bangs 2007, with permission from El Paso Water Utilities, El Paso, TX)**

*References*

Hutchison, William R., 2007. El Paso Groundwater Desalination Project: Initial Operation. Water Reuse and Desalination, As Bright as the Florida Sun, 2007 WateReuse Association Annual Symposium Proceedings.

Hutchison, William R., 2006. Desalination of Brackish Groundwater and Deep Well Injection of Concentrate in El Paso, Texas. In: Stars of the Future, Reuse & Desalination, 2006 WateReuse Association Annual Symposium Proceedings.

## ASCE/EWRI Task Committee

## CONCENTRATE MANAGEMENT IN DESALINATION

### Case Study

Management Approach: Deep-Well Injection

Committee Member(s): Berrin Tansel

Project Contact(s): Paul Mattausch, Howard Brogdon, North Collier County Regional Water Treatment Plant

Project Name: Concentrate Management by Deep-Well Injection at North Collier County Regional Water Treatment Plant in Naples, FL

Project Location: The administrative offices for the project are located at 3301 E. Tamiami Tr., Bldg H, Naples, FL

Desalination Process: NF (softening) and RO (desalination), NF and RO are two separate process, the permeate from both treatment processes are blended before degasification.

WTP Information:

- Rated Capacity: 20 MGD
- Max. Concentrate Flow: 4.8 MGD
- Typical Production: 17.6 MGD
- Typ. Concentrate Flow: 2.4 MGD

*Abstract*

Groundwater is the primary source for drinking water production in Collier County, Florida. The NCRWTP uses both nanofiltration (NF) and reverse osmosis (RO) treatment processes to produce drinking water for residential, commercial, industrial, and institutional uses. Groundwater is the sole source of raw water for drinking water production. The treatment plant is capable of treating 24.8 million gallons per day (MGD) with recovery rates of 85% and 75% for the NF and RO treatment trains, respectively, for an overall water production rate of 20 MGD. Pretreatment consists of sulfuric acid addition and cartridge filtration; high pressure pumps then feed the pretreated water into the NF train (approximately 120 psi) or the RO train (approximately 250 psi). Membrane concentrate is pumped into a deep well injection system consisting of two (2) class I deep injection wells. Well No. 1 is 3,300 ft deep and Well No. 2 which is 3,200 ft. The existing steel injection tubing was retrofitted with fiberglass (FRP) injection tubing and the annulus was filled with cement. The modification of the well design also changed the maximum injection rate and the operating pressure of the injection wells.

## *Process Design and Configuration*

Figure 1 displays a schematic diagram of the treatment process in the NCRWTP. Raw water is pretreated with sulfuric acid to adjust pH and to maintain carbonates in their soluable form along with a scale inhibitor to prevent other forms of scale. Pretreated water then passes through 5 micron cartridge filters to remove particulates from the raw water. The pretreated and filtered water then goes to the nanofiltration trains (fresh water treatment) or the reverse osmosis trains (brackish water treatment). Treated water (filtration permeate) is collected and piped to packed tower degasification for hydrogen sulfide removal. Sodium hydroxide is added to adjust the pH before the water goes to potable water storage and distribution to the public water supply system. The concentrate from the filtration processes is pumped into the deep injection well system.

The treatment plant is capable of treating 24.8 million gallons per day (MGD) with recovery rates of 85% and 75% for the NF and RO treatment trains, respectively, for an overall water production rate of 20 MGD. High pressure pumps for the NF trains and the RO trains are approximately 120 and 250 psi, respectively. Membrane concentrate is pumped into a deep well injection system consisting of two Class I deep injection wells. Well No. 1 (IW-1) is 3,330 ft deep and Well No. 2 (IW-2) is 3,210 ft deep. Maximum concentrate flow is designed at 6.3 MGD and normally operated at a flow rate of 2.4 MGD. The schematic diagram of the concentrate piping system is shown in Figure 2. Between the two deep injection wells is a dual-zone monitoring well.

## *Project background:*

The North Collier County Regional Water Treatment Plant is one of two water treatment plants in the Collier County Public Water Supply System, which provides service to over 170,000 permanent residents and over 200,000 residents in season, in portions of Collier County from Barefoot Beach to the Isles of Capri. The service area covers approximately 240 square miles, and water is distributed through nearly 900 miles of water main. In 1999, a facility expansion was completed that added 8 MGD of reverse osmosis water treatment process capacity to an existing 14 MGD NF treatment system, bringing the total constructed capacity of the facility to 20 MGD. A future expansion is planned to add an additional 2 MGD of high pressure reverse osmosis capacity to treat near-seawater quality groundwater, bringing the total constructed capacity of the North County Regional Water Treatment Plant to 22 MGD.

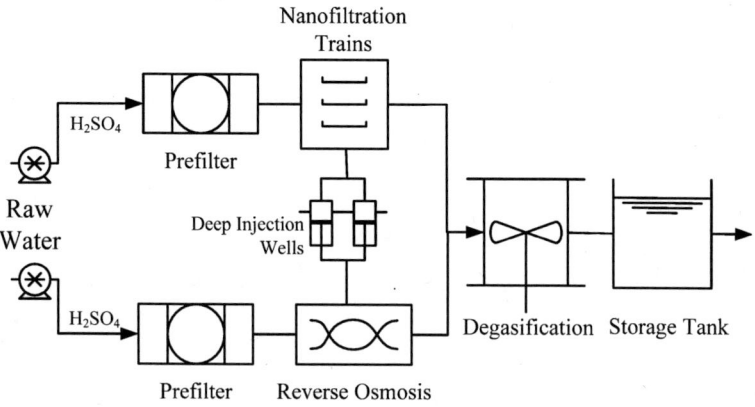

Figure 1. Schematic diagram of treatment process of NCRWTP (the raw water source for NF is fresh water, and brackish water is the source for RO) – (Courtesy of Berrin Tansel and Linda Lee)

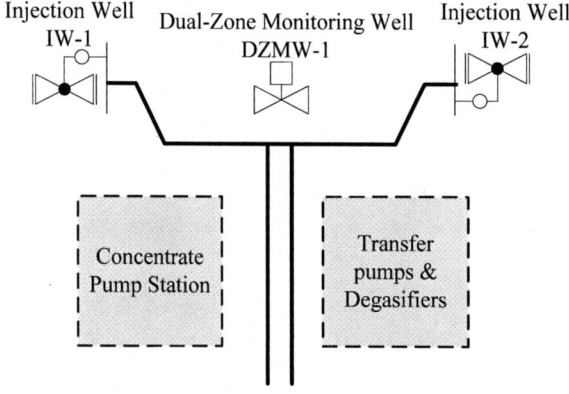

Figure 2. Schematic diagram of concentrate piping system (Courtesy of Berrin Tansel and Linda Lee)

### Description of the implemented solution:

The injection wells used at the NCCRWTP was originally designed with steel injection tubing. The NCCRWTP underwent a major modification for alternative well design. The existing steel injection tubing was retrofitted with fiberglass (FRP) injection tubing and also called for fully cementing the annulus. This also modified the pressure and maximum injection rate for operating the Injection Wells. Concentrate from the NF and RO systems is injected into the Oldsmar Formation at a maximum rate of 4,375 gallons per minute (gpm), or 6.3 million gallons per day (MGD). The maximum injection pressure was controlled at 100 pounds per square inch (psi).

### Data Collection Procedures:

Monitoring of the concentrate stream and its effect on the surrounding ground water ensures that the injection operation does not endanger the underground sources of drinking water through movement of the injectate or formation fluids. Samples are collected from a dual-zone monitoring well located between the two injection wells. Analysis of water from the monitoring wells and the injected concentrate stream include pH, temperature, conductivity, bicarbonate, chloride, calcium, magnesium, iron, potassium, sulfate, total dissolved solids, total Kjeldahl nitrogen, sodium, gross alpha and Radium 226/228.

### Permitting and regulatory overview and procedure:

The injection well permit was issued under the provisions of Chapter 403, Florida Statutes (F.S.), and Florida Administrative Code (F.A.C.) Rules 62-4, 62-520, 62-528, 62-550, and 62-660. The Application to Construct/Operate/Abandon Class I, III, or V Injection well System, DEP Form 62-528.900 (1), was received on September 19, 2008, with supporting documents and additional information last received on April 10, 2009. The demonstration of Financial Responsibility was complete as of April 13, 2009 and remains in effect.

### Analyses of plant concentrate:

The concentrate stream and its effect on the ground water surrounding the injection well are closely monitored by staff and laboratory personnel. Results from each analysis are reported to the Florida Department of Environmental Protection. Table 1 presents the quality of the concentrate in November 2009 (Collier County Government, 2010).

**Table 1. Concentrate stream and dual zone monitoring well water quality**[a]

| Parameter | Unit | Concentrate | Deep Injection Zone | Shallow Injection Zone |
|---|---|---|---|---|
| Bicarbonate Alkalinity | mg/L as $CaCO_3$ | 39.4 | 146 | 137 |
| Calcium | mg/L | 600 | 640 | 155 |
| Carbonate Alkalinity | mg/L as $CaCO_3$ | 5 | 5 | 5 |
| Chloride | mg/L | 2070 | 19600 | 2050 |
| Conductivity | µmhos/cm | 10700 | 49600 | 7640 |
| Gross alpha radioactivity | pCi/L | 26.5±5.91 | 38.5±8.15 | N.A. |
| Iron | mg/L | 0.62 | 1.7 | 3.2 |
| Magnesium | mg/L | 269 | 940 | 161 |
| Nitrogen Ammonia | mg/L as N | 1.2 | 0.17 | 0.54 |
| pH | S.U. | 5.7 | 7.6 | 7.9 |
| Potassium | mg/L | 78 | 384 | 55.5 |
| Radium 226 | pCi/L | 11.5±2.92 | 16.0±3.57 | N.A. |
| Radium 228 | pCi/L | 0.742±0.397 | 0.602±0.305 | N.A. |
| Sodium | mg/L | 1710 | 9900 | 1280 |
| Sulfate | mg/L | 2160 | 2340 | 628 |
| Temperature | °C | 29.3 | 30.3 | 27.4 |
| Total dissolved solids | mg/L | 7760 | 33300 | 5220 |
| Total Kjeldahl nitrogen | mg/L as N | 2.1 | 0.31 | 0.48 |

[a] as Reported in November, 2009

## *Economic Evaluation:*

The capital costs for the water treatment plant was about 6,000,000 dollars. The annual O&M costs are presented in the following table:

**Table 2. Annual O&M Costs**

| Annual O&M | Cost |
|---|---|
| Electric Power | $30,000 |
| Personnel | $2,000 |
| Parts, Chemicals | $1,000 |
| Miscellaneous | $12,000 |
| Total | $45,000 |

## *Key project lessons learned:*

The greatest potential risk to public health associated with deep injection wells in South Florida is vertical migration of wastewater, containing pathogenic microorganisms and pollutants, into brackish-water aquifer zones that are being used for alternative water-supply projects such as aquifer storage and recovery (Maliva et al, 2007). Based on the experience in the NCRWTP, no contamination of a groundwater source has been discovered. As discussed in Missimer's study (2009), from the TDS measurement in IW-2, the 10,000 mg/L TDS occurs at about 1175 ft below land surface, which suggests that higher-salinity water is still far away from the underground sources of drinking water for Collier County. These results demonstrate that the NCRWTP deep injection well system has been properly constructed, maintained and operated and the system is a successful example of a deep well injection concentrate disposal technique.

**Pictures**

(Photo by Collier County staff, with permission from Paul Mattausch, Director, Water Department, Collier County)

*Acknowledgment*

The authors greatly acknowledge the hospitality of the North Collier County Regional Water Treatment Plant working group, especially thanks for Mr. Howard Brogdon and Paul Mattausch for being very supportive of this case study.

*References*

Collier County Government (2010), Personnel Communication with Paul Mattausch, Director, Water Department, Public Utilities Division, Naples, FL.

North_Collier_County_WTP_50581_543_UO1I_Final_report

North County Regional Water Treatment Plant
http://www.colliergov.net/Index.aspx?page=627,
http://www2.hawaii.edu/~nabil/naples.htm

## ASCE/EWRI Task Committee
## CONCENTRATE MANAGEMENT IN DESALINATION
### Case Study

Management Approach: Deep Well Injection

Committee Member(s): Berrin Tansel

Project Contact(s):  Jay Ameno, Metcalf & Eddy, Inc., 3740 Executive Way Miramar FL 33025, Ph: 954 450 7770, Fax: 954 450 5100

Johanna Londono, Florida International University Ph: 305 348 2824, Fax: 305 348 2802

Project Name: Central Plantation Water Treatment Plant, Plantation, Florida

Project Location: 700 NW 91st Ave, Plantation, FL 33324

Desalination Process: Nanofiltration of brackish groundwater

WTP Information:

- Rated Capacity: 12 MGD
- Max. Concentrate Flow: 2.4 MGD
- Typical Production: 5 to 6 MGD
- Typical Concentrate Flow: 0.75 MGD

### *Process Design and Configuration*

Water is extracted from the Biscayne aquifer through 8 wells with depths ranging from 80 to 110 feet. Scale inhibitor is added to the raw water prior to cartridge filtration (5 micron), The water is then pumped through the NF membrane system at an average feed pressure of 130 psi. Each 2-stage array consists of fiberglass pressure vessels which contain the thin film composite NF spiral-wound membranes. In the first stage, the recovery rate of permeate is approximately 60% and in the second stage it is approximately 50% (of the rejected 40% from the first stage), yielding an average total recovery rate between 80 and 85%. The concentrate is returned to the boulder zone of the aquifer system to a depth of approximately 3,100 feet. The Boulder Zone begins at 2,790 feet (851 m) in depth and consists of a very hard, fractured, cavernous, dark dolomite.

After desalination, the permeate passes through a gas-stripping process to remove hydrogen sulfide and carbon dioxide to eliminate odor problems and control pH. Chlorine and ammonia are added for disinfection purposes. Fluoride is also added to the permeate. The mixing of these chemicals takes place in baffled tanks. After disinfection, water is stored before it is sent to the distribution system. This treatment plant is interconnected with the East Plantation WTP in order to supply drinking water to approximately 88,500 residents within its service area.

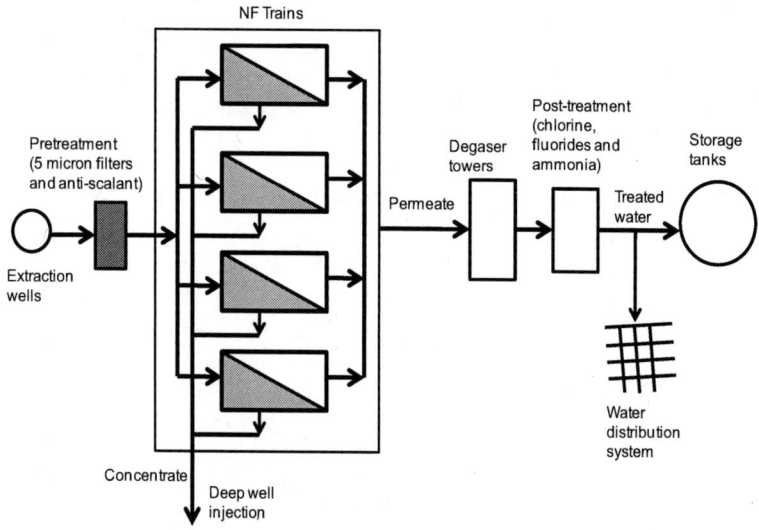

**Figure 1: Schematic of Pilot Treatment System (Courtesy of Berrin Tansel)**

## *Project background:*

In an effort to comply with the high levels of trihalomethanes and color, as well as to supply a higher quality of drinking water, the City of Plantation's conventional softening plant was modified and a NF plant was installed in 1991. Since then, the Central Plantation WTP has worked in conjunction with the East WTP to meet the community's potable water demand.

## *Description of the proposed solution:*

A Class I injection well was built according to the FDEP guidelines (described subsequently). The concentrate is returned to the boulder zone of the aquifer system through a single pipeline to a depth of approximately 3,100 feet. Continuous sampling and monitoring is carried out in the pipeline structure and in the monitoring wells for quality assurance. This system has been in operation since 1991. The material of the injection pipeline consists of PVC with concrete and steel coats.

As an alternative for the management of the concentrate, a pilot plant with RO treatment was built but it was not cost-effective for concentrate treatment purposes.

### *Data Collection Procedures:*

Monthly measurements are conducted and samples are collected in the surrounding monitoring wells, as well as in the deep well discharge, to ensure compliance with the regulations. Also, every 2.5 years, a mechanical integrity test (MIT) is conducted throughout the longitude of the injection pipe, to ensure its structural stability. Analytical reports are sent to the municipality and to FDEP.

### *Permitting and regulatory overview and procedure:*

Florida Department of Environmental Protection (FDEP) requires a comprehensive ground water quality protection program that regulates discharges to ground water. The program establishes ground water quality standards and classifications and permitting criteria. Within several FDEP rules there are construction and operation requirements, minimum setbacks, and ground water monitoring criteria. Dischargers to ground water are required to submit and implement a ground water monitoring plan with the monitoring wells, and a ground water sampling and analysis protocol. At a minimum, these plans require three monitoring wells: a background well, an intermediate well, and a compliance well. These wells are generally sampled quarterly, and the analysis is submitted to FDEP to ensure compliance with Florida's ground water standards.

Injection wells are required to be constructed, maintained, and operated so that the injected fluid remains in the injection zone, and the unapproved interchange of water between aquifers is prohibited. Class I injection wells are monitored so that if migration of injection fluids were to occur it would be detected before reaching the Underground Sources of Drinking Water (USDW). Testing is conducted on all Class I injection wells at a minimum of every five years to determine that the well structure has integrity.

According to Florida Administrative Code (FAC) 62-528, all Class I wells shall report monthly the following operating reports to the Department on:

> a. The physical, chemical, and other relevant characteristics of injection fluids;
> b. Daily readings of the pressure and flow for each well. For each domestic effluent disposal well, a specific injectivity test shall be performed quarterly;

c. Monthly average, maximum and minimum values for injection pressure, flow rate and volume, and annular pressure; and
d. The results of monitoring prescribed under Rule 62-528.425(1)(e), FAC.

### *Analyses of plant concentrate:*

There are no guidelines for reject water except that it must not contain hazardous waste constituents, as prohibited by the 62-528, FAC. The concentrate disposal well at the plant is considered a Class I well.

### *Economic Evaluation:*

Total plant construction cost $16 million in 1991 and the cost of treatment is approximately $0.92 cents/kgal. One considerable operating cost is the Mechanical Integrity Test of the well (every 2.5 years) due to the specialized work and equipment that is required. The UIC program requires applicants for Class I wells to assure, through a performance bond or other appropriate means, that resources necessary to cover post-closure monitoring and that any corrective action resulting from this monitoring are available [s. 62-528.435(9) of the UIC rule]. Guidance on fulfilling the financial responsibility requirements is set forth in the document, "Financial Responsibility Options for Owners and Operators of Injection Wells."

### *Key project lessons learned:*

There was significant corrosion of the initial steel injection pipeline. This problem was solved by changing the pipe material to PVC with a concrete and steel coating. Following this, FDEP increased the frequency of the Mechanical Integrity Tests, increasing the plant operation costs.

In general, the key constraints in the planning of the DWI system were:

- There must be a sufficiently receptive injection zone available.
- The final casing material are typically fiberglass reinforced polyester (FRP), but the packer hangers are still steel and so are exposed to corrosion; however, the FRP liner can be cemented in place to avoid the need for hangers or packers. It is recommended to consider the use of stainless steel as much as possible at the surface though it is costly and welds are a specialty.

The concept of building a single well with two monitoring zones is being reconsidered instead of two single monitoring zone wells. There have been several system failures of deep wells in Florida, and these can result in expensive (~$700K) rehabilitation efforts. A dual zone well is significantly less expensive to construct.

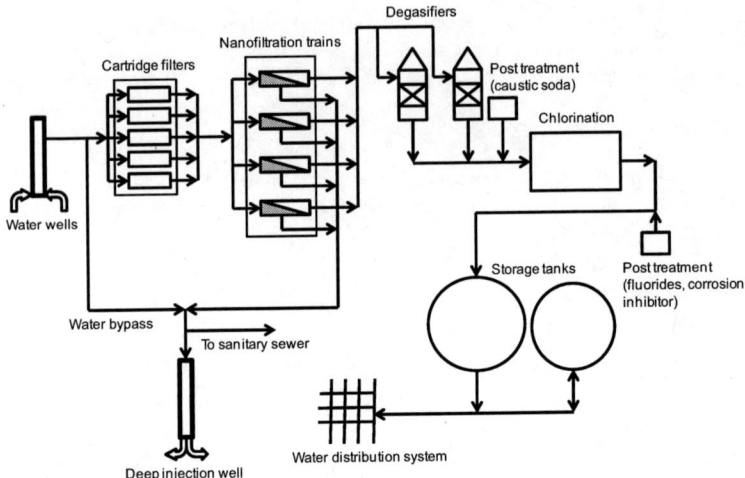

**Figure 2: Overall Schematic of Central Water Treatment Plant (Courtesy of Berrin Tansel)**

*References*

Licata, C.; Wengrenovich, M.; Breitenkam, H.. Developing an alternative water supply strategy for the city of Plantation, Florida. Proceedings of WEFTEC 2009, 3697-3712, October 11-14, Orlando, FL.

Truesdal, J.; Mickley, M.; Hamilton, R. Survey of membrane drinking water plant disposal methods. Desalination 102 (1995) 93-105.

2008 Integrated Water Quality Assessment for Florida. Florida Department of Environmental Protection. October 2008.

Chapter 62-528, Florida Administrative Code. Underground Injection Control.

American Desalting Plants Database. http://www2.hawaii.edu/~nabil/planp-2.htm

Plantation Utilities Department. http://plantation.org/Utilities/water-treatment.html

Underground Injection Control Program. Florida Department of Environmental Protection. http://www.dep.state.fl.us/water/uic/index.htm

## ASCE/EWRI Task Committee

## CONCENTRATE MANAGEMENT IN DESALINATION

### Case Study

Management Approach: Evaporation Ponds for disposal of desalination concentrate

Committee Member(s): James Jensen

Project Contact(s): Thomas Mannhardt / Ian Cameron[1]_Terry Fagg[2]__[1]Parsons Brinckerhoff, Brisbane Australia, [2]Western Downs Regional Council, Utilities Treatment Manager

Project Name: Dalby Stage 2 Desalination Plant

Project Location: Township of Dalby - Queensland, Australia

Desalination Process: Reverse Osmosis (MF/UF Pretreatment)

WTP Information:

- Rated Capacity: 1.1 MGD (4.0 MLD)
- Max. Concentrate Flow: 0.35 MGD (1.3 MLD)
- Typical Production: 1 MGD (3.6 MLD)
- Typ. Concentrate Flow: 0.32 MGD (1.2 MLD)

### *Abstract*

Decreased reliability of surface water and water quality issues with increased salinity resulted in the Dalby Town Council initiating a stage 2 desalination plant to supplement the existing water supply including an initial brackish water RO train installed in 2004 with a capacity of 0.45 MGD/1.7 MLD operating at a 75% recovery.

A Stage 2 desalination plant with two trains each a capacity of 0.52 MGD/2.0 MLD was proposed to be primarily supplied by coal seam methane (CSM) sourced water from approximately 100 extraction bore holes with a TDS level of approximately 6,000 mg/L was ultimately scrapped when an agreement could not be reached with the CSM operator. Instead, the original 0.45 MGD/1.7 MLD RO system was duplicated. Evaporation ponds are used for residuals management for the RO systems.

The ability to secure available land close to the treatment plant was a risk for the original RO project proceeding and critical for the management of the membrane concentrate. Reasonably accurate modeling input variables, including evaporation rate and the occurrence of storm events, needed to be determined to ensure the pond system was accurately sized.

*Process Design and Configuration*

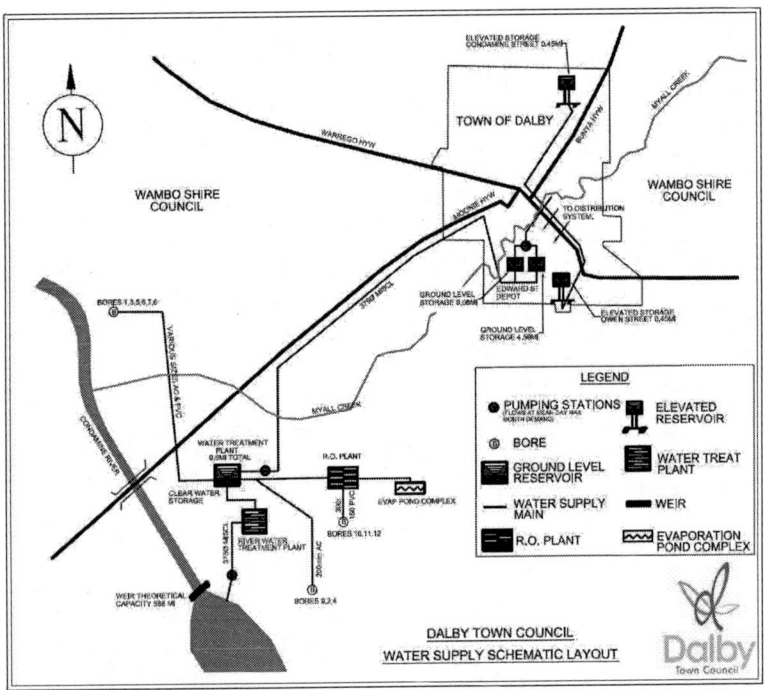

**Figure 1. Schematic of Dalby Water Supply System (Dalby Regional Council Drought Management Plan 2009, with permission from Western Downs Regional Council, formerly Dalby Town Council).**

### Project background:

The Dalby Stage 2 desalination project was initiated by the Dalby Regional Council to provide a sustainable water supply for the township of ~12,000 persons. The initial brackish water RO train with a capacity of 0.45 MGD/1.7 MLD was installed in 2004. The source water for the Stage 1 RO system is brackish groundwater with a TDS of approximately 2140 mg/L.

The Stage 2 project as originally proposed, consisted of two brackish water RO trains each with a capacity of 0.52MGD/2.0 MLD to be supplied by coal seam methane (CSM) sourced water from approximately 100 extraction bore holes with a TDS level of approximately 6,000 mg/L.. It was proposed that the concentrate stream from the 1st stage RO with a TDS of about 8,500 mg/L be used as the feed stream for the 2nd stage RO once installed, so as to reduce the total flow of concentrate to the evaporation ponds.

The Stage 2 desalination plant brought on line in 2010 has the same design as the Stage 1 plant with a capacity of 0.45 MGD/1.7 MLD operating at a 75% recovery. The water supply for the Stage 2 system is brackish groundwater with TDS of approximately 2,200 mg/L.

### Description of the proposed solution:

Due to the inland location and high evaporation rates, no suitable alternatives for disposal of the RO concentrate were available.

Evaporation ponds are used for residuals management for the RO systems. The ponds are designed to handle the projected rainfall for this specific region (100 year storm), and include an external embankment for potential flooding which is common for this area. The evaporation ponds are sealed with gypsum stabilized highly plastic silty clay found on site to prevent leakage of the saline concentrate into the ground water table. However, over the past 2 years the ponds have been retrofitted with a 2mm HDPE liner as the integrity of the clay liner could no longer be guaranteed. Monitoring of the groundwater had detected negative trends which although not proven indicated leakage out of the evaporation ponds. The Council decided that such a risk was unacceptable and progressively lined the ponds.

The availability of land located close to the RO system was a critical consideration. Twenty-one hectares of land from an adjoining property was acquired for the RO evaporation pond complex. Two of the four 2.5 m (8 foot) deep compartments are 7 ha (17.3 acres) in area and two are 3.5 ha (8.6 acres) in area (Walford).

***Data Collection Procedures:***

Monitoring of the ponds and associated infrastructure is conducted daily and comprises visual inspection of pipe work, monitoring of evaporation rates, salinity testing and monitoring of nearby test bores.

***Permitting and regulatory overview and procedure:***

The Australian EPA requires disposal of concentrate via approved means, in this case evaporation ponds. The ponds are required to be lined to prevent leakage into the ground water and to be suitably designed for the climatic conditions in that region. Critical design parameters are evaporation rates, rainfall, and flooding potential. The concentrate will ultimately form a salt-like cake; it was anticipated that removal and disposal to land fill will occur after 20 years of operation.

***Analyses of plant concentrate:***

No specific guideline or requirement for the analysis of the concentrate exists for the facility. The total dissolved solids content of the concentrate for the Stage 2 facility is estimated to the 16,500 mg/L.

***Economic Evaluation:***

Ultimately the salt-cake crystals that form in the bottom of the evaporation ponds will need to be removed; this is anticipated to occur in 20 years. The cost of this disposal will need to be factored into the operational cost of the plant over its life-cycle. This is difficult to predict as an approved site for safe disposal for this waste 20 years from now time has not yet been determined.

***Key project lessons learned:***

The ability to secure available land close to the treatment plant is critical for management of the membrane concentrate. This may take a long time and can present a significant risk to the project proceeding. Reasonably accurate modeling input variables, including evaporation rate and the occurrence of storm events, need to be determined to ensure the pond is accurate sized; over-sizing the pond can dramatically increase the cost of these types of projects.

**Pictures**

Figure 2. Lined evaporation ponds (Dalby Regional Council Drought Management Plan 2009, with permission from Western Downs Regional Council, formerly Dalby Town Council).

*Reference*

Dalby Regional Council March 2009. Drought Management Plan prepared by Brandon & Associates, Queensland.

Dalby Town Council. Dalby Water Supply Desalination Plant Information Guide: http://www.wdrc.qld.gov.au/services/documents/water/desalination_plant.pdf

Walford, Brian. Evaporative Lagoons for Dalby Desalination Project: Design and Construction Challenges. PB Network, Issue 64 Vol XXI No 3. December 2006: http://www.pbworld.com/news_events/publications/network/issue_64/64_09_Walford_evaporative.asp

# Appendix A-4
# Zero Liquid Discharge (ZLD) and Near ZLD

Water Desalination Concentrate Management and Piloting Study - South Florida – Sandeep Sethi

Closed-Loop Water Recovery and Recycling for Space Applications - NASA, Kennedy Space Center, FL – Berrin Tansel

## ASCE/EWRI Task Committee
## CONCENTRATE MANAGEMENT IN DESALINATION
### Case Study

Management Approach: Zero (or near-zero) Liquid Discharge Concentrate Disposal Systems

Committee Member(s): Sandeep Sethi, Carollo Engineers

Project Contact(s): Sandeep Sethi, Lyle Munce, David MacNevin (Carollo Engineers), Ashie Akpoji (South Florida Water Management District), Jeff Huren An (City of North Miami Beach)

Project Name: Water Desalination Concentrate Management and Piloting Study

Project Location: Florida, USA

Desalination Process: Reverse Osmosis (full-scale RO plant concentrate treated with RO pilot plant)

WTP Information:

- Rated Capacity: 6.0 mgd
- Max. Concentrate Flow: 2.0 mgd
- Typical Production: 4.0 mgd
- Typ. Concentrate Flow: 1.3 mgd

*Abstract*

The study included desktop evaluations of several reverse osmosis (RO) plants in South Florida Water Management District (SFWMD) and pilot testing at one representative plant. The pilot testing was undertaken at the City of North Miami Beach's Norwood Water Treatment Plant (WTP). The primary RO system operates at a recovery of 75 percent. Concentrate minimization will increase the net process recovery and make additional product water available. A dual RO system with intermediate chemical precipitation was selected for pilot testing of concentrate minimization.

## Process Design and Configuration

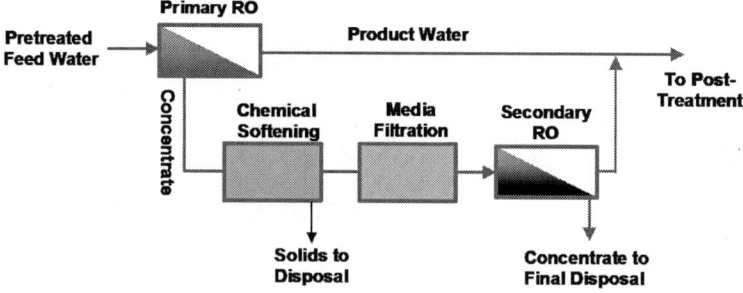

Figure 1. Schematic of Pilot Treatment System (Courtesy of Sandeep Sethi)

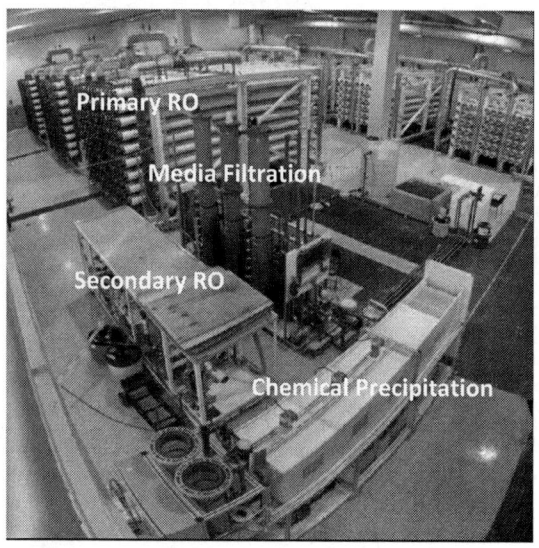

Figure 2. Layout of the Pilot Equipment (Photograph by Sandeep Sethi 2009, with permission from City of North Miami Beach Utilities, FL)

*Project background:*

Several public water utilities in South Florida have turned to brackish water desalination with RO in response to increasing demand for potable water and maximization of traditional fresh sources of water. The associated challenges with concentrate disposal can limit the use of alternative water sources in inland communities. Currently, the desalination capacity in the SFWMD is 206 mgd with a projected capacity of 540 mgd by 2025 (Akpoji, 2008). Increasing recovery efficiencies from the current average of 75 percent to about 95 percent will make over 100 mgd of water available by waste minimization.

The SFWMD undertook this study to evaluate alternatives for concentrate minimization in South Florida. The overall goal of the study is to provide recommendations for concentrate minimization through identification of affordable and sustainable treatment technologies.

*Description of the proposed solution:*

The City of North Miami Beach's WTP contains multiple processes for the treatment of fresh water and brackish water. A 15-mgd lime softening system treats fresh water from the Biscayne aquifer, a 9-mgd nanofiltration system treats highly organic freshwater from the Biscayne Aquifer water, and a 6-mgd reverse osmosis (RO) system treats brackish water from the Floridan Aquifer. Additionally, up to 1.5 mgd of raw Biscayne water and 0.5 mgd of raw Floridan water can be microfiltered and blended with the membrane permeate streams. The RO system operates at a recovery of 75 percent. The concentrate from the RO (and NF) is currently disposed via deep injection wells.

Dual RO with intermediate chemical precipitation is a physical-chemical approach to enhancing the recovery of a RO process through treatment and minimization of concentrate. This approach uses established technologies such as lime soda softening and a second phase RO (Sethi et al., 2009). The concentrate treatment step focuses on removal of cations of concern via precipitative softening to reduce the scaling potential of the concentrate. The steps involved are chemical treatment and precipitation for removal of calcium, magnesium, silica, and other sparingly soluble components, followed by filtration (e.g. media filtration or membrane filtration) for removing solids carryover from the precipitation process, and finally secondary RO. As the secondary RO system will be operated with higher TDS water, it will require higher pressures compared to the primary RO system.

Depending on the raw water chemistry, the secondary RO can recover an additional 50 to 60 percent or more of the primary RO concentrate as product water, resulting in a net recovery of greater than 88 percent.

## *Data Collection Procedures:*

Standard protocols for analytical sample collection and analysis were followed during the pilot testing, and the associated bench-testing that was performed immediately before the pilot testing to screen selected parameters for the pilot test. The pilot plant was operated for a period of three months during which samples were collected and analyzed at specific frequencies for all key streams: primary RO concentrate, post precipitation, post filtration, pretreated secondary RO feed, secondary RO permeate, and secondary RO concentrate. Parameters measured ranged from general water quality (TDS, TOC, turbidity, temperature, alkalinity, hardness etc) to specific inorganic, organic, and pathogens of concern. Most of the parameters were analyzed by an external laboratory while some parameters were analyzed in the field as needed (e.g. temparature, pressures, flows, conductivity, etc.). Selected paramteres were analyzed both in the field and external laboratory (e.g. pH, alkalinity, hardness).

## *Permitting and regulatory overview and procedure:*

The work conducted was a desktop and pilot study for assessment of concentrate minimization technologies. A preliminary assessment of the potential challenges to increased concentrate recovery for RO plants was performed as part of the study.

If concentrate recovery generates a solid or liquid hazardous waste, current concentrate disposal methods may be restricted or unavailable. An optimal approach to concentrate minimization would balance water efficiency with environmental health and safety.

In Florida, RO concentrate is classified as a non-hazardous "potable water byproduct" (Chapter 403.0882.(2) F.S.). Drinking water utilities are responsible to determine that they are not generating hazardous waste. The hazardous characteristic of concern for RO concentrate is toxicity. The Florida Department of Environmental Protection (FDEP) assesses toxicity using a "whole effluent toxicity" test (FDEP, 1995) that measures the aggregate toxicity of all substances in the waste stream (Chapter 62-4.241 F.A.C.). Waste streams discharging to surface waters must also meet surface water quality standards for specific contaminants (62-302 F.A.C.).

In addition to liquid wastes, concentrate treatment technologies may generate solid waste byproducts with potentially hazardous contaminant levels. For example, trace naturally occurring compounds in the Floridan Aquifer, such as radionuclides, arsenic, or others may accumulate to hazardous levels in solids from thermal brine crystallization, or lime-soda softening sludge. Florida has regulations prescribing special handling and landfill disposal requirements for hazardous wastes (Chapter 62-730 F.A.C.).

### *Analyses of plant concentrate:*

During the pilot study, the concentrate from the full-scale primary RO system demonstrated a TDS of approximately 11,000 mg/L. The concentrate from the pilot-scale secondary RO system demonstrated a TDS level of approximately 26,500 mg/L.

### *Economic Evaluation:*

Implementing concentrate minimization at the Norwood WTP using dual RO with intermediate precipitation is estimated to have a construction cost of about $11 million for treating 2-mgd of primary RO concentrate. At least 1-mgd of additional water would be generated through the concentrate minimization treatment.

While the brine concentrator alone will provide a much higher recovery (greater than 90 percent of the 2-mgd concentrate will be recovered) compared to the dual RO process, the total cost of treatment will be approximately double that of the dual RO process.

### *Key project lessons learned:*

The approach of dual RO with intermediate chemical precipitation for concentrate minimization demonstrated stable operation for a Florida brackish water with a moderate TDS level, and was shown to be conceptually viable through the pilot testing.

## Pictures

Figure 3. Equipment for intermediate chemical precipitation employed at the pilot testing at City of North Miami Beach Norwood WTP (Photograph by Sandeep Sethi 2009, with permission from City of North Miami Beach Utilities, FL)

Figure 4. Equipment for media filtration employed at the pilot testing at City of North Miami Beach Norwood WTP (Photograph by Sandeep Sethi 2009, with permission from City of North Miami Beach Utilities, FL)

Figure 5. Equipment for secondary RO employed at the pilot testing at City of North Miami Beach Norwood WTP (Photograph by Sandeep Sethi 2009, with permission from City of North Miami Beach Utilities, FL)

*Acknowledgements*

This project was undertaken and funded by the South Florida Water Management District, and the City of North Miami Beach sponsored the pilot site and provided associated support for the pilot testing.

*Reference*

Akpoji, A. (2008) Desalination in the South Florida Water Management District. Water Supply Department Internal Document. Pulled February 12, 2009. https://my.sfwmd.gov/pls/portal/docs/page/pg_grp_sfwmd_watersupply/subtabs - water conservatio - brackish/tab1610173/desalcombined-july08.pdf.

Carollo Engineers. (2009) Water Desalination Concentrate Management and Piloting, Report Prepared by Carollo Engineers for South Florida Water Management District, December 2009.

Sethi, S., Walker, S., Xu, P., and Drewes, J. (2009) Desalination Product Water Recovery and Concentrate Volume Minimization (Project # 3030). Report Number 91240, Water Research Foundation (formerly Awwa Research Foundation), Denver, CO.

## ASCE/EWRI Task Committee
## CONCENTRATE MANAGEMENT IN DESALINATION
### Case Study

Management Approach: Zero Liquid Discharge Concentrate Disposal Systems

Committee Member(s):   Berrin Tansel

Project Contact(s):   Berrin Tansel
  Phone: (305) 348-2928, Fax: (305) 348-2928

Project Name: Closed-Loop Water Recovery and Recycling for Space Applications

Project Location: NASA, Kennedy Space Center, Cape Canaveral, FL, USA

Desalination Process: Reverse osmosis of brackish groundwater

WTP Information (Bench scale):

- Rated Capacity: 6 L/day (Bench scale)
- Max. Concentrate Flow: 12 L/day
- Typical Production: 6 L/day
- Typical Concentrate Flow: 3 L/day (Bench scale)

### *Abstract*

One of the water recovery and recycling processes currently being evaluated for water recovery and recycling during space missions incorporates the aerobic rotational membrane system (ARMS) which was tested at the Kennedy Space Center, Florida. Membrane filtration can be used for downstream treatment of the effluents from biological treatment processes. However, the use of a membrane filtration process for bioreactor effluents requires both treatability and operational compatibility assessment to ensure that final effluent meets the required water quality standards with minimum or no significant fouling within the anticipated operational conditions. A series of filtration tests were performed both with distilled water to establish a baseline permeability of the NF and RO membranes and followed by ARMS effluent.

### *Process Design and Configuration*

Water recovery and recycling is an important consideration for reducing payload during long-term space missions. The aerobic rotational membrane system (ARMS) was developed and operated at the Kennedy Space Center (KSC) for water recovery and recycling during long space missions. A closed loop water recovery and recycling system, to improve the wastewater quality to drinking water quality, requires integration of multiple treatment technologies. The experimental studies have shown that microfiltration did not provide significant quality improvement since the bioreactor effluent contained primarily dissolved salts (Tansel, et al., 2005). Therefore, to improve quality of the bioreactor effluent, it was necessary to use nanofiltration (NF) or reverse osmosis (RO) systems. The concentrate from the membrane filtration step contained about 90% of the solids that were present in the ARMS bioreactor effluent. Evaporation was one of the possible technologies which was considered to further recover water from the concentrate.

## *Project background:*

It is estimated that the water needs in space would be about 11.5 L/day-person. Although this quantity is less than 4% of the water used on Earth per person per day, it would not be possible to transport the amount of water needed to space during long missions. Hence, a compact closed loop water recovery and recycling system is needed for long space missions. One of the water recovery and recycling processes currently being evaluated for water recovery and recycling during space missions incorporates the aerobic rotational membrane system (ARMS) which was tested at the Kennedy Space Center, Florida. The ARMS is a novel compact membrane bioreactor which converts ammonia to nitrates. The 12-L bioreactor is operated in continuous mode with simulated wastewater representing the output of 1/2 person crew during space missions. It is operated at a hydraulic retention time of 48 hours, with a throughput of 6 L/day (Garland et al., 2003). The ARMS is aerated through the rotating hallow fiber membrane module consisting of silastic hollow fibers. Oxygen is provided through the rotating hallow fiber membrane module to the biofilm on the membranes. The rotating hallow fiber membrane module provides mixing to maintain high mass transfer rates between the bulk liquid and the biofilm (Rector et al., 2004a and 2004b) resulting into high bioconversion rates which reduce the volume requirements of the system. However, the bioreactor effluent contains significant amounts of dissolved inorganic salts as well as bacterial products which need to be removed for development of a closed loop water recovery and recycling system.

## *Description of the proposed solution:*

Membrane filtration can be used for downstream treatment of the effluents from biological treatment processes. However, the use of a membrane filtration process for bioreactor effluents requires both treatability and operational compatibility assessment to ensure that final effluent meets the required water quality standards with minimum or no significant fouling within the anticipated operational conditions. The following figures were developed by Berrin Tansel.

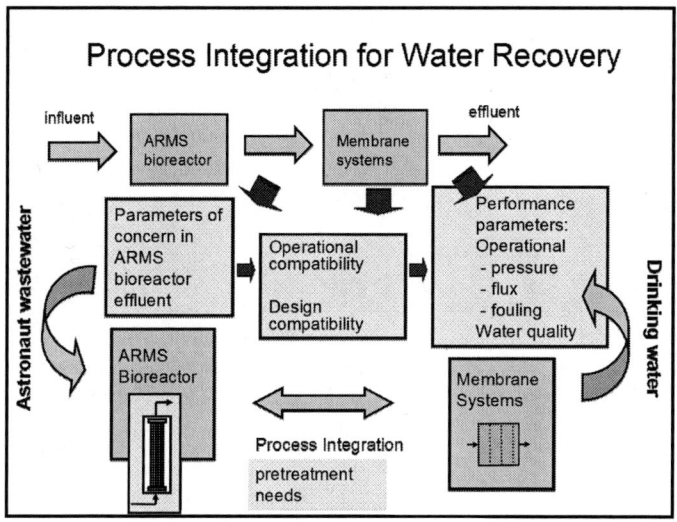

The quality of the effluent from the bioreactor was improved by sequential filtration using NF and RO membranes.

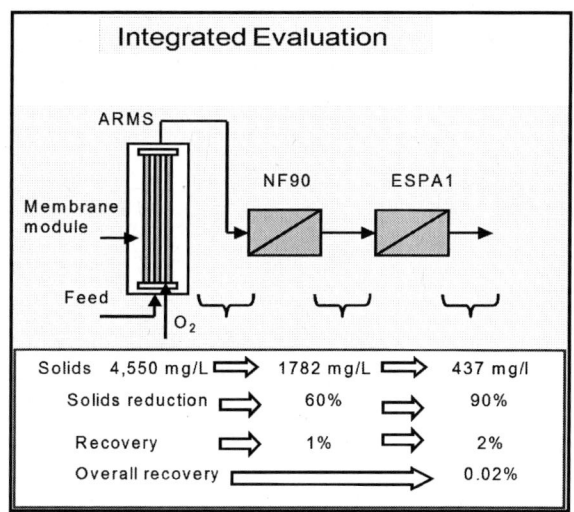

Soft deposits were observed on the membrane surface during inspection. Progression of the membrane fouling was assessed by analysis of the images of the membrane captured by atomic force microscopy (AFM). Analysis of the images of the membranes indicated that deposition of extracellular polymeric substances (EPS) on the membrane surface progressively increased until a critical thickness was achieved. After 4-day use, the surface morphology of the membrane appeared to be significantly different from that of the clean membrane.

### *Data Collection Procedures:*

A series of filtration tests were performed both with distilled water to establish a baseline permeability of the NF and RO membranes and followed by ARMS effluent. The permeate samples were analyzed for total solids according to Standard Methods. The ions were analyzed by the Dionex DX-120 ion chromatograph.

### *Permitting and regulatory overview and procedure:*

The work conducted was a feasibility study for technical assessment of coupling biological treatment processes with membrane filtration as a downstream treatment step.

### *Analyses of plant concentrate:*

NA

### *Economic Evaluation*

NA

*Picture (Courtesy of Berrin Tansel)*

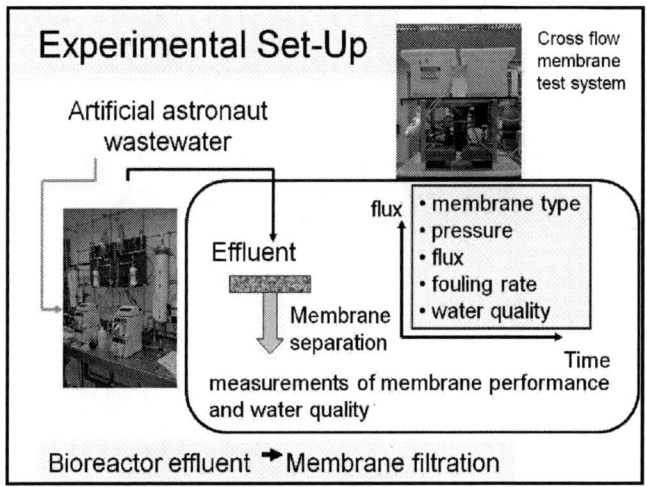

## References

Rector, T.J., Strayer, R.F., Hummerick, M.P., Garland, J.L., Roberts, M.S., Levine, L.H., 2004a, Performance Evaluation of a Submerged Bioreactor for Lunar Applications, Engineering Construction and Operations in Challenging Environments Earth and Science, Proceedings of the Ninth Biennial ASCE Aerospace Division International Conference, League City, Houston, TX.

Rector, T., Garland, J., Srayer, R.F., Levine, L., Roberts, M., Hummerick, M., 2004b, "Design and Preliminary Evaluation of a Novel Gravity Independent Rotating Biological Membrane Reactor," SAE International, 34th International Conference on Environmental Systems (ICES), July 19-22, 2004, Colorado Springs, CO, Paper No. 041CES-160.

Tansel, B.; Sager, J.; Rector, T.; Garland, J.; Strayer, R.F.; Levine, L.; Roberts, M.; Hummerick, M.; Bauer, J., 2005, Integrated Evaluation of a Sequential Membrane Filtration System for Recovery of Bioreactor Effluent During Long Space Missions, Journal of Membrane Science, 255(1-2), pp. 117-124.

# Index

Page numbers followed by *t* indicate a table, those followed by *f* indicate a figure, and those followed by *e* indicate an equation.

allocated impact zones 24–25
Australia case study, evaporation pond 111–115, 112*f*, 115*f*

brackish water 6–8, 6*f*, 10–11, 35

California case studies: discharge to ocean 60–65, 63*f*, 64*f*, 65*f*; municipal water district 50–59, 51*f*, 55*t*, 56*t*, 58*f*; plant collocation seawater 44–49, 45*f*, 46*f*
Carlsbad Seawater Desalination Plant (proposed) case study 44–49, 45*f*, 46*f*
cartridge filtration 8
Central Plantation (FL) Water Treatment Plant case study 106–110, 107*f*, 110*f*
Charles Meyer Desalination Facility case study 60–65, 63*f*, 64*f*, 65*f*
chemical softening 10
Closed-Loop Water Recovery and Recycling for Space Applications case study 125–129, 127*f*, 129*f*
concentrate 5, 6; management and disposal methods 9–13; stream characteristics 8–9
costs. *See* economic evaluation

Dalby Stage 2 Desalination Plant (Australia) case study 111–115, 112*f*, 115*f*
deep well injection 9, 12; environmental issues 19, 23, 27–28
deep well injection, case studies: Central Plantation Water Treatment Plant 106–110, 107*f*, 110*f*; Kay Bailey Hutchison Desalination Plant 90–98, 90*f*, 97*f*, 98*f*; North Collier County Regional Water Treatment Plant 99–105, 101*f*, 103*t*, 104*t*, 105*f*
demand management, economic evaluation 36–37
drinking water treatment plants (DWTPs) 17–18

economic evaluation 35; National Research Council recommendations 35–37; Southern California Technologies summary 37; typical summary 37–38, 38*t*
electrodialysis (ED)/electrodialysis reversal (EDR) processes 5–6
energy costs 35, 36, 37
environmental impact statement (EIS) 19
environmental issues 21; assessments and permits 22–23; concentrate characteristics and membrane fouling 21–22, 22*t*; deep well injection 19, 23, 27–28; evaporation ponds 29–30; land application 22, 28–29; oceans and bays 23–25; regulation and 19; sanitary sewers 22, 26–27; surface water 25–26; zero liquid discharge systems 30
Environmental Protection Agency (EPA). *See* National Pollution Discharge Eliminations System (NPDES)
evaporation ponds 9, 10, 12, 13; environmental issues 29–30
evaporation ponds, case study: Dalby Stage 2 Desalination Plant (Australia) 111–115, 112*f*, 115*f*
Florida 19, 23, 24, 25, 28, 29
Florida case studies: deep well injection, Naples 99–105, 101*f*, 103*t*, 104*t*, 105*f*; deep well injection, Plantation 106–110, 107*f*, 110*f*; low pressure reverse osmosis 73–78, 76*f*, 77*f*, 78*f*; reverse osmosis 68–72, 72*f*; zero liquid discharge, NASA 125–129, 127*f*, 129*f*; zero liquid discharge, South Florida 118–124, 119*f*, 123*f*, 124*f*
forward osmosis (FO) 10
fouling, of membranes 7–8, 7*f*

hollow fine fiber RO elements, 7

injection zones. *See* deep well injection

Joe Mullins Reverse Osmosis Water Treatment Facility case study 68–72, 72f

Kay Bailey Hutchison Desalination Plant case study 90–98, 90f, 97f, 98f

land application 9, 12; environmental issues 22, 28–29

Marin Municipal Water District case study 50–59, 51f, 55t, 56t, 58f
marine ecosystems. *See* oceans and bays
membrane desalination 5–8, 6f
membrane distillation (MD) 10

nanofiltration (NF) process 5–6
NASA, Closed-Loop Water Recovery and Recycling case study 125–129, 127f, 129f
National Environmental Policy Act (NEPA) 19
National Pollution Discharge Eliminations System (NPDES) 11, 17–18, 26, 28
North Collier County Regional Water Treatment Plant case study 99–105, 101f, 103t, 104t, 105f

oceans and bays 10–11; environmental issues 23–25
oceans and bays, case studies: Carlsbad Seawater Desalination Plant (proposed) 44–49, 45f, 46f; Charles Meyer Desalination Facility 60–65, 63f, 64f, 65f; Marin Municipal Water District 50–59, 51f, 55t, 56t, 58f
Ormond Beach WTP Low Pressure Reverse Osmosis (LPRO) Expansion case study 73–78, 76f, 77f, 78f

permeate 6
pilot testing, minimization methods 13–14
Pilot-Research Membrane Treatment of Non-Irrigation Season Flows case study 79–88, 80f, 83t, 84t, 85f, 86f, 87f, 88f
process and configurations 5; concentrate management and disposal methods 9–13; concentrate stream characteristics 8–9; membrane desalination processes 5–8, 6f; pilot testing of minimization methods 13–14
publicly owned treatment works (POTWs): environmental issues 23, 26; NPDES permits 18

recovery, of water 8–9, 9e
regulation 17; environmental impacts and 19; National Pollution Discharge Eliminations System 17–18; solid wastes 19; underground injection 18; U.S. federal agencies 17
rejection, of pollutants 9e
reprocessing 10
reverse osmosis (RO) process 5–6; economic evaluation 35, 36

Safe Water Drinking Act (SWRA) 18
sanitary sewers and surface water 9, 10–11; environmental issues 22, 25–27
sanitary sewers and surface water, case studies: Joe Mullins Reverse Osmosis Water Treatment Facility case study 68–72, 72f; Ormond Beach WTP Low Pressure Reverse Osmosis (LPRO) Expansion case study 73–78, 76f, 77f, 78f; Pilot-Research Membrane Treatment of Non-Irrigation Season Flows case study 79–88, 80f, 83t, 84t, 85f, 86f, 87f, 88f
Santa Ana Regional Interceptor (SARI) 11
scaling, of membranes 7–8, 7f
seawater desalination 8, 10, 11. *See also* oceans and bays
sewers. *See* sanitary sewers and surface water
solar ponds 12, 29
solid waste regulation 19
Southern California Technologies 37
spiral wound membrane elements 7
surface water. *See* sanitary sewers and surface water

Texas, 28
Texas case studies: deep well injection 90–98, 90f, 97f, 98f; membrane treatment of season flows 79–88, 80f, 83t, 84t, 85f, 86f, 87f, 88f
thermal evaporators. *See* evaporation ponds
Total Dissolved Solids (TDS) 5, 8

underground injection control (UIC) 18

Water Desalination Concentrate Management and Piloting Study 118–124, 119f, 123f, 124f
whole effluent toxicity (WET) 25

zero liquid discharge systems 10, 30

zero liquid discharge systems, case studies:
  Closed-Loop Water Recovery and Recycling for Space Applications  125–129, 127*f*, 129*f*;

Water Desalination Concentrate Management and Piloting Study  118–124, 119*f*, 123*f*, 124*f*